Gardener's Latin

a Lexicon by Bill Neal

Introduction by Barbara Damrosch

Robert Hale · London

ISBN 0-7090-5106-9

8 10 12 14 15 13 11 9

Robert Hale Limited
Clerkenwell House
Clerkenwell Green
London EC1R OHT

Printed in Singapore by
Kyodo Printing Co (S'pore) Pte Ltd

Algonquin Books of Chapel Hill acknowledges with thanks permission to quote short passages from the following books: *Adventures with Hardy Bulbs*, by Louise Beebe Wilder, © 1989, Macmillan Publishing Co.; Selection from *The Alice B. Toklas Cookbook* by Alice B. Toklas, © 1954 by Alice B. Toklas; Copyright © renewed 1982 by Edward M. Burns. Reprinted by permission of HarperCollins Publishers Inc.; *Color in My Garden*, by Louise Beebe Wilder, copyright © 1990, by Harrison Wilder Taylor, used with permission of Atlantic Monthly Press; *The Garden in Autumn*, by Allen Lacy, © 1990, used with permission of Atlantic Monthly Press; *Green Thoughts*, by Eleanor Pereyni, © 1981, Random House, Inc.; *The Little Bulbs*, by Elizabeth Lawrence, © 1986, Duke University Press, reprinted with permission of the publisher; *A Rock Garden in the South*, by Elizabeth Lawrence © 1990, Duke University Press, reprinted with permission of the publisher; *Southern Gardens, Southern Gardening*, by William Hunt, © 1982 Duke University Press, reprinted with permission of the author; *Through the Garden Gate*, by Elizabeth Lawrence and edited by Bill Neal (Chapel Hill: The University of North Carolina Press, 1990), reprinted by permission of the *Charlotte Observer*; *The Treasury of Flowers*, by Alice M. Coats, © 1975, Phaidon Press for the Royal Horticultural Society, London; *The Illustrated Garden Book*, by V. Sackville-West, © 1986, by Nigel Nicolson, reprinted with permission of Atheneum, an imprint of Macmillan Publishing Co.; and *Wild Flowers of North Carolina*, by Williams S. Justice and C. Ritchie Bell (Chapel Hill: The University of North Carolina Press, 1987).

"Speaks Latin, that satin doll."

—*Billy Strayhorn and Johnny Mercer*

or a dead language, Latin gets a lot of use these days, especially by gardeners. Friends who used to grow "loosestrife" and "bee balm" now talk about their *Lythrum* and their *Monarda,* oblivious to the scorn of people like my neighbor Ellen. "My flowers grow fine without knowing what genus they belong to, and what's good enough for them is good enough for me," Ellen insists. But for every Ellen there is someone who can't sleep until she knows whether her catmint is *Nepeta faassenii* or *Nepeta mussinii.*

Snobbery? Pretension? I think not. Gardening has grown in popularity, so more gardeners are concerned about doing it well. Often this means knowing with more certainty which plants are which and about plant histories, native habitats, and growing requirements. And sooner or later this comes down to "speaking Latin."

Common plant names are a rich trove of imagery, and I would never suggest that we stop talking about pigweed, pussytoes, or love-in-a-puff. But common names can be troublesome when it is time to go shopping. Let's say, for example, that your mother wants a "rose of Sharon" for her birthday. This name is commonly applied to two plants, one *Hypericum calycinum,* a foot-high groundcover that blooms in midsummer, another *Hibiscus syriacus,* a shrub up to fifteen feet tall that blooms at summer's end— neither of them roses. By relying solely on the common name you risk buying one plant when you thought you were choosing the other.

Finding your way in the jungles of botanical Latin can teach you a great deal about the plants themselves. Take "baby's breath," a poetic name that beautifully describes the plant's airy cloud of tiny white blossoms. By also learning its Latin name, *Gypsophila paniculata,* you will better remember that the plant grows best in calcareous soil (*Gypsophila* means "gypsum lover") and the flowers are arranged in panicles (loose, spreading flower clusters). Another species, *Gypsophila repens* is more low growing, since *repens* means "creeping." In the end, learning botanical names will make you more than just a better shopper; it will make you a better gardener and increase your understanding of the natural world.

Does this mean that you have to learn the entire Latin language before you can pick up a spade and go out into the garden? Not at all. Botanical nomenclature is a highly organized, practical system that uses a limited and highly specialized vocabulary—a little language of its own, really. You need not memorize it, just learn how it works and begin to feel at home in it. As you run across certain words again and again they start to become familiar.

We owe this nomenclature to Carl Linnaeus, a Swedish naturalist who in the mid-eighteenth century designed a system for classifying and naming plants, animals, and minerals that is still used today. During the seventeenth century others had begun to organize plants into large groups based on structural similarities, and into smaller groups called genera (singular, genus) within those. Linnaeus broke these genera down into 7,300 smaller units, or species, and came up with a system of plant names based on the genus and the species to which each belonged. In so doing Linnaeus helped make evident the natural order of plants. It is an order based on structural features that we now know evolved over the centuries in order to insure the plants' survival. He decoded, in effect, an order that was already there in nature. And he helped make it desirable, even fashionable, to study and grow interesting plants by giving the world the first universally understood system for naming them.

By naming plants in Latin Linnaeus was following the accepted practice of his day.

Latin was not only the written but also the spoken language of the educated—the one in which people from different countries could communicate. Since it was a language still in use, the Latin of Linnaeus's time had become quite different from the classical Latin of ancient Rome. In being adapted to botanical purposes it underwent still other changes by incorporating words from Greek and other tongues, and by adapting to the ever-growing knowledge of plants. The classical world knew relatively little about plants and the way they behaved. (See margin note for *Cinnamomifolius* on page 30 for a comical example.) Many plant parts were completely unknown before the invention of magnification glasses in the seventeenth century, especially sexual parts; the sexuality of plants was not discovered until shortly before the time of Linnaeus.

With these new demands being made upon it, botanical Latin became the little language unto itself, and a richly descriptive one at that. Where a word did not exist in Latin or Greek, a comparison was usually made. In order to name, Linnaeus defined; in order to define, he described; in order to describe, he created metaphors. Hence botanical terms often seem as poetic as a string of Homeric epithets.

Plant names consist of two words, the genus name and the species name. Before Linnaeus, plants were identified only by lengthy recorded descriptions that varied from author to author. His system of two-word, or "binomial," nomenclature provided an agreed-upon way of naming that was brief, concise, and at the same time informative. The genus name (capitalized) was a noun in the nominative singular, the species (in lowercase) an adjective describing the noun and agreeing with it in case and number. Hence in the name *Acer palmatum*, the tree we call Japanese maple, the neuter noun for maple, *Acer*, is described by the adjective *palmatum*, which means "shaped like a hand" (referring to the shape of the leaves). Sometimes there is a third Latin name denoting a subspecies or naturally occuring variety, as in *Acer palmatum dissectum*—a Japanese maple whose leaves are very finely dissected, like threads.

Of special importance to the gardener are horticultural varieties, proliferated not by nature at large but by the hand of man. A capitalized name in single quotations following the genus and species indicates a "cultivar" produced by human hybridization or selection. Hence *Acer palmatum dissectum* 'Flavescens' is a threadleaf Japanese maple bred or selected for its yellowish leaves.

Sometimes plant sellers skip over the species name and give just the genus and the horticultural variety. This is too bad because the species name often tells you most about the plant, I, for one, am always struggling to track down the species from which a garden hybrid is derived, in hopes of finding out whether it will, say, grow tall, bloom late, or like its feet to be wet. Fortunately there has been such a surge of interest in growing simple original species that species names are coming back into currency.

As new plant species are discovered they are named and thereby join the body of our plant knowledge. How do you know when you have found a new species? According to the great twentieth-century botanist L. H. Bailey,

> It is impossible accurately to define what is meant by a species. The naturalist gradually acquires the idea and it becomes an unconscious part of his attitude toward living things. Nature is not laid out in formal lines. Perhaps it will aid the inquirer if I repeat the brief definition I wrote in Hortus: A kind of plant or animal that is distinct from other kinds in marked or essential features, that has good characteristics of identification, and that may be assumed to represent in nature a continuing succession of individuals from generation to generation.

Someone who discovers a new species can name it anything he or she wants, within the very precise rules of the International Code of Nomenclature. Hence the well-circulated tale about the wag who named a new organism *fungusamongus* is probably apocryphal. (A name should be "real" Latin unless it derives from a proper name.) But it is a true fact that the bacterium thought to cause cat scratch fever, *Afipia felix*, was named after AFIP—The Armed Forces Institute of Pathology. And mischievous scientists have been known to give foul-smelling fungi the names of eminent mycologists.

The pages that follow are a glossary (that is, a partial dictionary) of a large number of botanical names you are likely to encounter, listing species names only, in the masculine singular form. Bill Neal's margin notes offer some of the charming lore and literary associations that the names have acquired down through the centuries.

The species names as defined are no less charming. A plant that is *tipuliformis* is shaped like a daddy longlegs; an *arachnoides* plant is cobwebby; a *blepharophyllus* plant is one "fringed like eyelashes"; and the one called *pardalianthes* "has the strength to strangle a leopard." By paying attention to species names we can sometimes learn a plant's origin (*californicus*, *transylvanicus*), that its favorite habitat is a salt marsh (*salsuginosus*), or that it blooms only at night (*noctiflorus*). We might learn that it feels prickly to the touch (*aculeatissimus*) or smells like a goat (*hircinus*). Knowing that words ending in *-folius* describe the leaves, *-florus* the flowers, *-caulis* the stem, or *-carus* the fruit will increase your understanding of them.

How will a new plant behave in your garden? If it is *compactus* it will probably stay small, or if *columnaris* make a vertical accent. If *flore-pleno* it will have numerous or many-petaled flowers; if *sempervirens* it is evergreen, at least in some climates. If it is *admirabilis* or *elegantissimus* it might delight you to grow it. If *horridus*, *fatuus*, or *phu* you might be sorry.

So, are Latin plant names dry, absurdly polysyllabic nonsense spouted only by bumptious, long-winded overachievers? They are only if you don't know what they mean.

—BARBARA DAMROSCH

bbrevia'tus: abbreviated or shortened; shorter than adjoining parts

abieti'nus: resembling fir trees

aborti'vus: aborted; exhibiting arrested development; defective

abrotanifo'lius: with leaves resembling the finely cut foliage of southernwood, *Artemisia abrotanum*

abrup'tus: abrupt; coming to a sudden termination

absinthoi'des: resembling wormwood, *Artemisia absinthium*

abyssin'icus: Abyssinian; Ethiopian

acadien'sis: of or from Nova Scotia, Canada; Acadian

acanthifo'lius: with broad, ornamental, spiked foliage like that of *Acanthus*

acantho'comus: having spiny hairs

acau'lis: stemless; apparently stemless; having the stem underground

ac'colus: dwelling nearby; neighboring

aceph'alus: headless; the absence of a head, as in certain leafy cabbages such as kale

acer'bus: harsh or sour-tasting; bitter

acerifo'lius: with leaves resembling the maple, *Acer*

aceroi'des: maple-like

acero'sus: needle-shaped and rigid; like the pine's leaves

A COTTAGE NOSEGAY
Abrotanifolius. "*Artemisia Abrotanum.* Southernwood. Lad's Love. Old Man. A low growing hoary evergreen shrub with richly aromatic feathery foliage. Found often in old gardens it once played a part in every cottager's nosegay. Few sweeter are to be devised than Southernwood and white Moss Rosebuds."
Louise Beebe Wilder,
The Fragrant Path, 1990

CORINTHIAN ORDER ↑
Acanthifolius. Acanthus leaves crown one of the most ornate and easily recognized elements of ancient architecture, the Corinthian capital.

SALVE OF ACHILLES ↑
Achilleaefolius. "I want you to cut out this arrow from my thigh, wash off the blood with warm water and spread soothing ointment on the wound. They say you have some excellent prescriptions that you learnt from Achilles."
Homer, *Iliad*
"This plant Achillea is thought to be the very same, wherewith Achilles cured the wounds of his soldiers."
John Gerard, *Herbal*, 1597

achilleaefo'lius: Achillea-leaved; having finely cut, ferny foliage like that of the yarrow

acicula'ris: needle-like; narrow, stiff, and pointed

acina'ceus: scimitar-shaped; like a short, crescent-shaped sword

acinacifo'lius: scimitar-leaved

acinacifor'mis: scimitar-shaped

aconitifo'lius: having leaves like the aconite; palmately cleft

acrot'riche: hairy-lipped

aculeatis'simus: very prickly

aculea'tus: prickly; set about with prickles, as a fruit or seed

acumina'tus: sharpened; tapering to a point

acu'tus: acute; sharp-pointed

adenoph'orus: bearing or producing glands, usually referring to sticky, nectar-bearing glands

adenophyl'lus: having glands on the leaves

adiantoi'des: with leaves like the maidenhair fern, *Adiantum*

admira'bilis: admirable; noteworthy

adna'tus: adnate, or grown together; with attached surfaces (usually used to indicate two dissimilar structures growing together naturally but in an apparently abnormal position)

adonidifo'lius: with leaves like *Adonis,* rather fennel-like

adpres'sus: pressed against; lying flat against, as the hairs on the stems of some plants *adscen'dens:* ascending; rising gradually

adsur'gens: ascending; rising to an erect position; straight up

adun'cus: hooked; like the beak of a parrot; crooked; bent

ad'venus: newly arrived; adventive; added to but not essential; not native

aegypti'acus: Egyptian; of or from the valley of the Nile

aem'ulus: emulating; rivaling

aene'us: bronze- or copper-colored

aequinoctia'lis: pertaining to the equinox; mid-tropical; having flowers that open or close at regular intervals

aequipet'alus: equal-petaled

aequitri'lobus: having three equal lobes

ae'rius: aerial; above ground or water

aerugino'sus: bluish green, like verdigris; like the color of oxidized copper

aestiva'lis: pertaining to summer

aesti'vus: summer

aethiop'icus: Ethiopian; African, south of Libya and Egypt

aetol'icus: of or from Aetolia, Greece

affi'nis: related to, especially in terms of structure; allied to another species

a'fra: African

MOMENT'S PLEASURE ✓
Adonidifolius. The ancient world, from Syria to Greece, worshiped Adonis as the god of vegetation. A peculiar function of his worship was the planting of the Adonis garden. Wheat, barley, lettuces, and fennel seeds were planted in earthen pots and baskets. These pots were grouped around a statue of Adonis on the roof. The plants shot up quickly, and in the intense rooftop heat they soon withered. After eight days the dead plants were thrown into the sea with the little image of the god. These little plants that sprang up so quickly and withered so soon were a symbol of the short life of the god. And so a garden of Adonis has come to mean a small and short-lived pleasure.

ADMIRABLE THINGS
Agavoides. "The moon-
light touching o'er a
terrace ∾ One tall
Agave above the lake."
Alfred, Lord Tennyson,
"Daisy," 1842
Tennyson rightly
admired the *Agave*, for
its Greek root is *agavos*,
literally, admirable.
But perhaps its great-
est admirers are the
Mexicans who use it
for, among other
things, making rope
and fiber, and as the
source of tequila
and pulque.

❦

PERENNIAL
CONFUSIONS
Ageratoides. "The
Coelestina ageratoides, a
half-hardy perennial
with blue ageratum-
like flowerheads,
much employed in
bedding, must not be
confounded with the
true Ageratums."
W. Thompson, *Trea-
sury of Botany*, 1866

africa'nus: African

agavoi'des: like the *Agaves* ↘

ageratifo'lius: with leaves like the common garden
  plant *Ageratum*

ageratoi'des: Ageratum-like

aggrega'tus: aggregate; clustered; applies to flowers or fruits
  collected into one mass

agra'rius: of the fields; growing wild in cultivated land

agres'tis: of or pertaining to fields or cultivated land

agrifo'lius: scabby-leaved

aizoi'des: resembling the *Aizoaceae*, a family of low-growing
  evergreen herbs with fleshy, often silvery foliage

ala'tus: winged; having wings or appendages
  that appear to be wings

albes'cens: whitish; becoming white

al'bicans: whitish

albicau'lis: white-stemmed

al'bidus: white

albiflo'rus: having white flowers

al'bifrons: white-fronded

albispi'nus: white-spined

albocinc'tus: white-girdled; white-crowned

albomacula'tus: having white spots

albopic'tus: white-painted

albopilo'sus: having white hairs

albospi'cus: white-spiked

al'bulus: whitish

al'bus: white

alcicor'nis: resembling the horns of the elk or moose

alep'picus: of or from Aleppo, Syria

alexandri'nus: of or from Alexandria, Egypt

al'gidus: cold; chilly

alien'us: foreign

allia'ceus: of *Allium*; garlic- or onion-like in shape, odor, or taste ↗

alliariaefo'lius: like the leaves of *Alliaria*; having kidney-shaped to heart-shaped, finely cut foliage

alnifo'lius: with leaves like the alder

aloi'des: *Aloe*-like; having pointed, erect, succulent leaves

aloifo'lius: with leaves like the *Aloe*

alopecurioi'des, alopecuroi'des: like *Alopecurus*, commonly called meadow foxtail, a grass that grows up to six feet with fluffy seed spikes

alpes'tris: nearly alpine; from just below the timberline

alpig'enus: alpine; found high in the mountains

alpi'nus: alpine

alta'icus: of the Altai Mountains, Asia

alter'nans: alternating, as opposed to opposite, plant parts

alternifo'lius: with leaves alternately spaced, not opposite

STINKING LILY

Alliaceus. Its detractors call it the stinking lily. But along with its relatives the onions and leeks, garlic—*Allium sativum*—is among the most ancient of cultivated plants. Both the Babylonians and the Chinese prescribed its use thousands of years before Christ. Herodotus says that the great pyramid of Cheops was inscribed with the amount spent on leeks and onions for the laborers. The Israelites in the desert paid the *Alliums* an immortal compliment. In the Bible they cried out, "We remember the fish we ate in Egypt for nothing, the cucumbers, the melons, the leeks, the onions, and the garlic; but now our strength is dried up, and there is nothing at all but this manna to look at."

alter'nus: alternating; alternate

althaeoi'des: Althaea-like; hollyhock-like

al'tifrons: having tall fronds

altis'simus: very tall; tallest

al'tus: tall

alum'nus: well-nourished, flourishing, or strong

alyssoi'des: Alyssum-like

amab'ilis: lovable; amiable; lovely

amaranthoi'des: like the plant *Amaranth*, often with brilliant
foliage and striking flower heads

amarantic'olor: Amaranth-colored; intensely reddish purple

amaricau'lis: bitter-stemmed

ama'rus: bitter

amazon'icus: of or from the Amazon River region

ambig'uus: ambiguous; of doubtful classification

ambly'odon: blunt-toothed

ambrosioi'des: of or like *Ambrosia* (named after the nectar of
the gods) a term applied to several different herbs in
the old days; in North America, a common garden
pest, ragweed

amelloi'des: of or like the Italian starwort; named after the
river Mellus near Rome; aster-like

america'nus: American

amethys'tinus: amethystine; violet-colored

amethystoglos'sus: amethyst-tongued

ammoph'ilus: sand-loving; growing in sandy places

amoe'nus: charming; pleasing

amphib'ius: amphibious; suited for or adapted to growing on land or in water

amplexicau'lis: stem-clasping, usually referring to a leaf whose base embraces the stem

amplexifo'lius: having leaves that clasp the stem

amplia'tus: enlarged; increased

amplis'simus: most or very ample

am'plus: ample; large

amuren'sis: of or from the Amur River region, the border of Manchuria and Siberia

amygdalifor'mis: shaped like an almond

amygdal'inus: almond-like

amygdaloi'des: almond-like

anacan'thus: without thorns or spines

anacardioi'des: like the fruit of the *Anacardium*, the cashew nut tree

anagyroi'des: bearing recurved pods like those of *Anagyris*, a Mediterranean shrub

anatol'icus: of or from Anatolia or Turkey, Asia Minor

an'ceps: two-headed; two-edged, as of a flattened, two-edged stem; also, problematic, equivocal, or of doubtful origin

ancyren'sis: of or from Ankara, Turkey

MEADOW STARS ↑
Amelloides. "There is also a meadow-flower that farmers call ∾ Amellus. You will find it easily, ∾ For from a single clump it pullulates ∾ With a mass of stems; the disc itself is golden, ∾ But in the abundant petals round about ∾ Crimson is shot with violet."
Virgil, *Georgics*

THE WIND'S TEARS ↑
Anemoneflorus. The
Anemone is commonly
called the windflower.
Turner wrote in his
Herbal: "Anemone hath
the name because the
floure never openeth it
selfe, but when the
wynde bloweth." Red
Anemones were thought
to take their color
from Adonis's blood,
white ones from
Aphrodite's tears.

andic'olus: native of the Andes

andi'nus: Andine, pertaining to the Andes

androg'ynus: hermaphrodite; having male and female
flowers on the same spike

androsa'ceus: like *Androsace,* a small rock-garden plant of the
primrose family, commonly called rock jasmine

anemoneflo'rus: with flowers like the *Anemone*

anemonefo'lius: with leaves like the *Anemone*

anethifo'lius: like the foliage of dill, *Anethum*

aneu'rus: nerveless

anfractuo'sus: twisted; winding; sinuous; spirally twisted

an'glicus: English; of England

angui'nus: snaky; snake-like

angula'ris, angulat'us: angular; angled

angulo'sus: angled; full of corners

angu'ria: like a cucumber

angustifo'lius: narrow-leaved

angus'tus: narrow

anisa'tum: anise-scented

anisodor'us: with the odor of anise

anisophyl'lus: exhibiting inequality in two leaves of a pair as
to shape or size

annot'inus: year-old; belonging to last year; denoting
distinct yearly growths

annular'is: annular; ring-shaped; arranged in a circle

annula'tus: marked with rings; surrounded by raised rings

an'nuus: annual; living only one year or one plant season

anom'alus: anomalous; out of the ordinary; incongruous

anopet'alus: with erect petals

antarc'ticus: of the antarctic regions

anthemoi'des: like *Anthemis;* resembling chamomile

anthocre'ne: with or like a flower fountain

anthyllidifo'lius: with leaves like *Anthyllis,* the kidney vetch or woundwort

antilla'ris, antille'ris: of or from the Antilles, West Indies

antip'odum: of or from the Antipodes

antiquo'rum: of the ancients

anti'quus: ancient

antirrhiniflo'rus: Antirrhinum-flowered; resembling the flowers of the snapdragon

antirrhinoi'des: Antirrhinum-like; snapdragon-like

apenni'nus: pertaining to the Apennines, Italy

aper'tus: uncovered, bare, or open

apet'alus: without petals

aphyl'lus: leafless

apicula'tus: apiculate; ending abruptly in a point

apif'era: bee-bearing

apiifo'lius: Apium-leaved; with leaves like those of the carrot family

ap'odus: without a foot-stalk; sessile

A CRUEL JOKER ↑
Apifera. Among nature's crueler jokes, *Ophrys apifera,* the bee orchid, plays the siren. The flower's labellum, decked out in bee disguise, drives the drones to sexual frenzy as they attempt copulation.

EAGLE OR DOVE ↑
Aquilegifolius. Columbines take their botanical name—*Aquilegia*—from the shape of the petals, which resemble an eagle. On the other hand, the common name means dove. The leaves are tripartite, attractive, and medicinal according to the herbals, "used in lotions with good success for sore mouths and throats." Nicholas Culpeper, *Culpeper's Complete Herbal,* 1813

apopet'alus: having free petals

appendicula'tus: appendaged; with small, often hanging appendages; in some mushrooms, when the veil appears cobwebby

applana'tus: flattened; horizontal

applica'tus: joined; attached

ap'ricus: uncovered; growing in open, sunny places

ap'terus: wingless

aquat'icus, aquat'ilis: aquatic; living in or under water

a'queus: watery

aquifo'lius: holly-leaved; with pointy leaves

aquilegifo'lius: *Aquilegia*-leaved; with leaves like the columbine

aquili'nus: aquiline; eagle-like

arab'icus: Arabian

arachnoi'des: spider-like; cobwebby; covered with long and scraggly hairs

araliaefo'lius: with deeply cut leaves like *Aralia* (ginseng and English ivy, for instance)

arbores'cens: becoming tree-like; woody

arbo'reus: tree-like

arbus'culus: like a small tree

arc'ticus: arctic

arcua'tus: bow-like; bent

arena'rius, areno'sus: of sand; growing in sandy places

areola'tus: pitted

argenta'tus: silvery; silvered

argenteogutta'tus: silver-spotted

argen'teus: silvery

argilla'ceus: of clay; growing in clay; clay-colored; drab

argophyl'lus: silver-leaved

argu'tus: sharp-toothed

argyrae'us: silvery

argyroc'omus: silver-haired

argyroneu'rus: silver-nerved

argyrophyl'lus: silver-leaved

ar'idus: arid; growing in dry places; withered; lacking sap

arieti'nus: like a ram's head; like the horns of the ram

arista'tus: aristate; bearded; with a bristle-like appendage

aristo'sus: bearded

arizon'icus: of Arizona, United States

arkansa'nus: of Arkansas, United States

arma'tus: armed; equipped; thorny

armeni'acus: blush-colored; of or pertaining to Armenia

armilla'ris: with a bracelet, arm-ring, or collar; encircled

aromat'icus: aromatic; fragrant

arrect'us: raised up; erect; pointing upward

artemisioi'des: Artemisia-like; like wormwood, usually with aromatic grayish green foliage

articula'tus: articulated; jointed

CULTIVATED AROMAS
Aromaticus. Cloves, among the most fragrant of spices, are dried flower buds from the tropical tree *Eugenia aromaticus.*

CHASTE WORMWOOD ↑
Artemisioides. Artemisia takes its name from Artemis, goddess of chastity. It is an herb of which Mrs. Wilder says diplomatically, "The Artemisias are a large family, all having odoriferous qualities, some nicer than others." Louise Beebe Wilder, *The Fragrant Path,* 1990

ROUGH EDGES
Asper. The rough leaves of *Doodia aspera* give it its graphic English name, hacksaw fern.

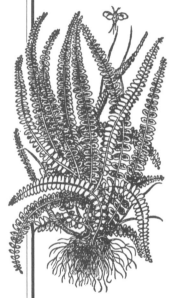

FOR MEN ONLY ↑
Asplenifolius. The spleenworts were used to treat ailments of the spleen and liver, but for men only. These plants were believed to cause barrenness in women.

arundina'ceus: resembling a reed or cane

arven'sis: pertaining to cultivated fields

arvonien'sis: of or from Caernarvonshire, North Wales, Britain

asarifo'lius: Asarum-leaved; resembling the heart or kidney-shaped leaves of the wild ginger

ascalon'icus: of or from Ascalon, Syria

ascen'dens: ascending

asclepiade'us: Asclepias-like; resembling the butterfly weed

asiat'icus: Asian

as'per: rough

aspericau'lis: rough-stemmed

asperifo'lius: rough-leaved

asper'rimus: very rough

asplenifo'lius: with foliage like that of the *Asplenium* ferns (spleenwort)

assim'ilis: similar; like to

assur'gens: ascending; rising upward

assurgentiflo'rus: flowers ascending

asteroi'des: aster-like; star-like

astu'ricus: of the Asturias, Spain

atamas'co: literally, "it is stained or streaked with red"

a'ter: coal black

atlan'ticus: of the Atlantic coast; of or from the Atlas Mountains, northwest Africa

atoma'rius: speckled

atra'tus: blackened

atriplicifo'lius: Atriplex-leaved; orach-leaved

atrocar'pus: dark-fruited

atropurpu'reus: dark purple

atro'rubens: dark red

atrosanguin'eus: literally, dark blood red

atroviola'ceus: dark violet

atro'virens: dark green

attenua'tus: attenuated; tapering gradually to a point

at'ticus: pertaining to Attica or Athens, Greece

aubretioi'des: resembling *Aubretia*, purple rock cress, a low-growing rock-garden plant of the Mediterranean

augu'rius: of the soothsayers

augustis'simus: most notable or majestic

augus'tus: notable; majestic; august

auranti'acus: orange-red

aurantifo'lius: golden-leaved

aure'olus: golden

au'reus: golden yellow

auric'omus: golden-haired

auricula'tus: eared; having ear-like appendages

auri'tus: eared

australien'sis: of or belonging to Australia

austra'lis: of the Southern Hemisphere

ANCIENT SPINACH ↑
Atriplicifolius. Atriplex or orach, a wild green, was eaten by the ancients like spinach. Culpeper says: "It is so commonly known to every housewife, it were labour lost to describe it." Today it is mostly consumed by Australian grazing sheep.

NAKED NANNIES
Autumnalis. "The Michaelmas crocus is the meadow saffron, *Colchicum autumnale.* The flowers are sometimes called 'naked nannies.' In my garden some of its varieties have bloomed at Michaelmas, and the white form nearly always does. . . . It is native to the English water meadows where it must be a fine sight if there are any such places left as those Bishop Mant described a hundred or more years ago: 'go to Monmouth's level meads, ∾ Where Wye the gentle Monmow weds; ∾ Long brilliant tubes of purple hue ∾ The ground in countless myriads strew.' " Elizabeth Lawrence, *Through the Garden Gate,* 1990

austri'acus: Austrian

austri'nus: of the Southern Hemisphere

autumna'lis: autumn; fall-blooming →

avicula'ris: pertaining to birds

a'vium: of the birds

axilla'ris: axillary; of or pertaining to the axils; the angle between a stem and leaf on the upper side

azaleoi'des: Azalea-like

azor'icus: of the Azores, Portugal

azu'reus: azure; sky blue

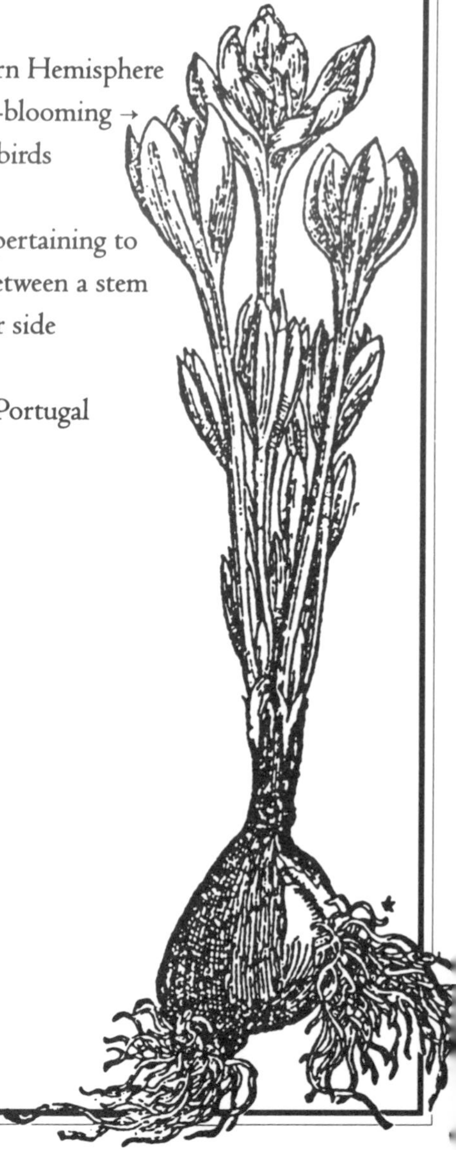

abylon'icus: Babylonian

bac'cans: bearing berries

bacca'tus: having fruits with a pulpy, berry-like texture

baccif'era: producing or having berries

bacterioph'ilus: bacteria-loving

baet'icus: of or from Spain (Baetica is the Latin name for Andalusia)

bahian'us: of or from Bahia, Brazil

baicalen'sis: from or near Lake Baikal, Siberia

balcan'icus: of or from the Balkans

balear'icus: of or from the Balearic Islands, Spain

balsa'meus: balsamic; yielding a fragrant gum or resin

balsamif'era: yielding a fragrant gum or resin

bal'ticus: of or from the Baltic Sea region

bambusoi'des: with leaves or growth like bamboo

banat'icus: of Banat, Romania

bar'barus: foreign; of or from the Barbary Coast, Africa

barbat'ulus: somewhat bearded

barba'tus: barbed; bearded; producing long, weak hairs

barbell'ata: fringed; having minute barbs

barbig'era: bearing barbs or beards

barbiner'vis: having bearded or barbed nerves

POCKET STRAWBERRIES ↓
Balsameus. Thoreau wrote that the young shoots of the balsam fir, *Abies balsamea,* when picked and kept in a pocket for a few days emit the fragrance of strawberries, "only it is somewhat more aromatic and spicy."

A BARBED SUGGESTION
Barbellata. The gentians are commonly without fragrance, but Louise Beebe Wilder found in her correspondence that the rare fringed gentian "has about the most delightful fragrance of any Colorado flower, not very strong but suggestive of Strawberries and other aromatic things." Louise Beebe Wilder, *The Fragrant Path,* 1932

DREADFUL PESTO
Basilicus. Basil as pesto is almost as popular as pasta these days, but traditionally basil was an herb of dread and suspicion. Culpeper wrote that it "is the herb which all authors are together by the ears about, and rail at one another, like lawyers. Galen and Dioscorides hold it not fitting to be taken inwardly, and Chrysippus rails at it with downright Billingsgate rhetoric." Culpeper continues: "Hilarius, a French physician, affirms upon his own knowledge, that an acquaintance of his, by common smelling to it, had a scorpion bred in his brain." He concludes: "I dare write no more of it." Nicholas Culpeper, *Culpeper's Complete Herbal,* 1813

❀

barbino'de: bearded at the nodes or joints

barbula'tus: small-bearded; somewhat or short-bearded

bartiseaefo'lius: with leaves like *Bartisia,* an herbaceous alpine plant

baselloi'des: like the climbing tropical plant *Basella*

basila'ris: pertaining to the base or bottom; basal

basil'icus: princely or royal

bata'tus: Carib Indian or Haitian name for the sweet potato

bavar'icus: Bavarian

belg'icus: of or from Belgium

belladon'na: beautiful lady

bellidifo'lius: having beautiful leaves; having leaves like the English daisy, *Bellis* →

bellidifor'mis: daisy-like; like *Bellis*

bellidioi'des: daisy-like; like *Bellis*

bel'lus: handsome; beautiful

benedic'tus: blessed; well spoken of

bermudian'us: of or from Bermuda

beta'ceus: beet-like

betonicaefo'lius: with leaves like the European mint, betony, *Betonica*

betulaefo'lius: with leaves like the birch, *Betula*

betuli'nus: birch-like

Bellis Spinosa Cretica.

betuloi'des: birch-like

bicarina'tus: having two keels (a keel is a ridge-shaped part of the flower resembling a boat's keel)

bic'olor: two-colored

bicor'nis, bicornu'tus: two-horned, as in heaths with two-horned anthers

bicort'icus: having two barks

bidenta'tus: with two teeth; doubly toothed as along an edge of a leaf

bien'nis: biennial; completing the life cycle in two growing seasons, usually blooming and fruiting in the second

bif'idus: twice cut; forked; having two segments

biflo'rus: having two flowers

bifo'lius: having or producing two leaves

bifor'mis: of two forms

bi'frons: two-fronded; exhibiting two faces or aspects

bifurca'tus: twice forked; having two prongs

bigib'bus: with two swellings or projections, usually basal

biglu'mis: having two glumes (husks or bracts); in grasses, having two empty bracts at the base of the spike

BEAUTIFUL LADY

Belladonna. Amaryllis belladonna is the true *Amaryllis;* the showy bulbs of the catalogues for winter-forcing are *Hippeastrums.* Carl Linnaeus called this native of the Cape of Good Hope "belladonna"—beautiful lady—for its attractive pink-and-white complexion.

A DAISY, I PRESUME

Bellidiformis. Dorotheanus bellidiformis was languishing as a weed in the Kirstenbosch Botanic Garden in Cape Town when Samuel Ryder, founder of the English seed firm of the same name, found it in 1928. Mr. Ryder asked for all the seed that could be supplied, renamed the flower Livingstone daisy after the explorer David Livingstone, and made it a popular success in his catalogue.

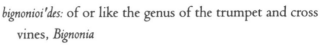

INDIAN CIGARS ↑
Bignonioides. The handsome *Catalpa* species native to the southern states with its broad, heart-shaped leaves is called *bignonioides* after the foliage of the *Bignonias,* the cross vines. Children call it "cigar tree" and sometimes smoke its long, slender fruits.

❦

HONEY BERRY
Bijugus. Melicoccus bijugatus is a tropical American tree grown for its sweet, juicy fruit. In markets it is offered as mamoncillo, Spanish lime, genip, and honey berry.

❦

bignonioi'des: of or like the genus of the trumpet and cross vines, *Bignonia*

bi'jugus: yoked; two together; doubly yoked; two pairs joined

bi'lobus: having two lobes

bina'tus: twin; in twos or pairs; bifoliate

binerva'tus, biner'vis: having two prominent nerves or veins

binocula'ris: two-eyed; two-spotted

biparti'tus: having two parts; divided nearly in two

bipet'alus: having two petals

bipinnatif'idus: twice pinnately cut; like a pinnate leaf whose sections are again pinnate (with leaflets on either side of a common stem in a feather-like arrangement)

bipuncta'tus: having two spots or dots

bisec'tus: cut in two parts

biserra'tus: twice toothed; having doubly serrate teeth, as on a leaf's margin

bispino'sus: having two spines

bistor'tus: twice twisted

bisulca'tus: having two grooves or furrows

biterna'tus: twice ternate; of a leaf, with three main divisions, each division with three leaflets

bitumino'sus: bituminous; coal black; tarry; adhesive

bival'vis: having two valves; of a fruit, splitting into two parts at maturity, as a pea pod *blan'dus:* mild; not bitter; pleasing; charming

blephariglot'tis: having a fringed tongue

blepharophyl'lus: with leaves fringed like eyelashes

bombycif'erum: silky

bona-nox: good-night; night-flowering

bonarien'sis: of or from Buenos Aires, Argentina

bononien'sis: of or from Bologna (formerly Bononia), Italy

bo'nus: good

borbon'icus: of or from Bourbonne, France; of or from
  Réunion Island (formerly Bourbon Island),
  Indian Ocean

borea'lis: northern

botryoi'des: resembling a bunch of grapes ↘

botulifor'mis: sausage-shaped

brachia'tus: branched at right angles

brachyan'drus: short-stamened

brachyan'thus: short-flowered

brachyb'otrys: short-clustered

brachycar'pus: short-fruited

brachypet'alus: short-petaled

brachyp'odus: short-stalked

brachyt'ylus: short-styled; short-knobbed

bractea'tus: having or bearing bracts (leaf-like and often
  highly colored organs just beneath, and sometimes
  eclipsing the flower)

NORTHERN
BOUQUETS
Borealis. "*Galium boreale*
is a pretty little native
that has long grown in
my garden in out-of-
the-way places and
contributes its modest
stems of creamy fra-
grant flowers to bou-
quets of scentless
blooms. This is called
the Northern Bed-
straw."
Louise Beebe Wilder,
The Fragrant Path, 1990

❦

FADED DWARF
Brachycarpus. The
dwarf evening primrose
is *Oenothera brachycarpa,*
native to Colorado,
with showy yellow
flowers that fade to
orange.

❦

SHORT AND STINKING
Brevifolius. Yucca brevifolia
is the eerie Joshua tree
of the western deserts,
blooming at night
from March to May
with stinking flowers.

✤

SUNSET GRASSES
Brizaeformis. Briza
is a hardy annual grass
from the Mediter-
ranean. It was a favorite
of E. A. Bowles and in
My Garden in Summer he
wrote, "Its pendant
heads are as lovely as
any green thing in the
garden in early June.
. . . Later on the heads
turn yellow, then buff,
and still later the
stalks ripen to a rich,
foxy burnt-sienna red,
and especially just
before sunset a large
patch lights up and
looks wonderfully
brilliant even among
the flaming colours
of a July border."

✤

bracteo'sus: having numerous or conspicuous bracts

brasilia'nus: Brazilian

brassicaefo'lius: resembling the leaves of *Brassica*—cabbage, mustard, or cress

brevicauda'tus: short-tailed

brevicau'lis: short-stemmed

brevifo'lius: short-leaved

brev'ifrons: short-fronded; short-leaved

brevipeduncula'tus: short-peduncled; short-stalked

brev'ipes: short-footed or short-stalked; with a short petiole

breviros'tris: short-beaked

brev'is: short

brevisca'pus: short-scaped

brevise'tus: short-bristled

brevi'spathus: short-spathed

brevis'simus: very short; shortest

brevi'stylus: short-styled

brilliantis'simus: very brilliant

brittan'icus: of or from Great Britain

brizaefor'mis: resembling *Briza*; the quaking grass †

bronchia'lis: bronchial; used to treat bronchitis

brumal'is: of the winter solstice; winter-flowering

brun'neus: deep brown

buc'arius: of or from Bokhara, Turkistan

buceph'alus: ox-headed

buddleifo'lius: having leaves like *Buddleia,* the butterfly bush

buddleoi'des: Buddleia-like

bufo'nius: pertaining to the toad; growing in damp places

bulbif'era: bulb-bearing

bulboco'dium: having a wooly bulb

bulbo'sus: bulbous

bulgar'icus: of or from Bulgaria

bulla'tus: with a blistered or puckered surface; having a rounded knob

bupleurifo'lius: with rounded leaves like *Bupleurum*

burman'icus: of or from Burma

buxifo'lius: with leaves like the box, *Buxus →*

byzanti'nus: Byzantine; of or from Istanbul, Turkey

NARCISSUS'S
NEW DRESS
Bulbocodium. "Narcissus
bulbocodium is regarded
as a relatively recent
species, still in the
process of evolution
—the narcissus, as
it were, trying on a
new dress. Neverthe-
less the 'novelty' prob-
ably dates back to the
time when Spain and
North Africa formed
a single land-mass."
Alice M. Coats,
The Treasury of Flowers,
1975

❧

PUCKERED MINT
Bullatus. Stachys bullata,
the western wood
mint, is a troublesome
weed in parts of Cali-
fornia and Oregon, but
the soft, wrinkled
foliage is aromatic
when bruised.

❧

WICKERWORKER
Calamifolius. Calamus is the largest genus of the palms with more than 300 tropical species. Many are climbing plants including *C. rotang*, one of the sources of rattan from which wicker-work is made.

DON'T BE FOOLED
Calendulaceus. "Perdita's 'Marigold, that goes to bed with the sun, and with him rises, weeping,' (Shakespeare) is not (as Burpee would have you think) the modern marigold *(Tagetes),* but *Calendula officinalis,* which has a long season, and so is usually in bloom for all the festivals of the Virgin."
Elizabeth Lawrence,
Through the Garden Gate,
1990

 aca'o: Aztec name for the cocoa tree, *Theobroma*

cachemir'icus: of Kashmir, south Asia

cad'micus: with a metallic appearance; like tin

caenos'us: growing in muddy places

caerules'cens: becoming or verging on dark blue

caeru'leus: cerulean; dark blue

cae'sius: bluish gray; light gray; lavender

caespito'sus: tufted; forming matted mounds like some grasses

caf'fer, caf'fra: of or from Kafir, South Africa

cajanifo'lius: with leaves like the tropical pigeon pea, *Cajanus,* called *cajan* in Malay

calab'ricus: from Calabria, Italy

calamifo'lius: with leaves like a reed; resembling *Calamus*

calathi'nus: with a basket-like flower head

calcara'tus: spurred

calca'reus: pertaining to lime; growing in chalky places; with the color and/or texture of chalk

calceola'tus: shaped like a shoe

calcif'ugus: disliking lime

calendula'ceus: like the European pot marigold, *Calendula* ↑

califor'nicus: of California, United States

caligino'sus: growing in misty places

callianth'us: having beautiful flowers

callicar'pus: bearing beautiful fruits

callistach'yus: having beautiful spikes

callizo'nus: having beautiful zones or bands

callo'sus: thick-skinned; with calluses; exhibiting a hardened or thickened surface

caloceph'alus: having a beautiful head

caloc'omus: having beautiful hair

caloph'rys: with dark margins

calophyl'lus: having beautiful leaves

cal'vus: bald; hairless; naked; glabrous

calyc'inus: calyx-like; with a persistent calyx

calycula'tus: resembling a small calyx

calyptra'tus: bearing a calyptra, a hood-like cover over fruit or flower

cambodgen'sis: of or from Cambodia

cambodien'sis: of or from Cambodia

cam'bricus: Cambrian; Welsh

camelliiflo'rus: with flowers like the *Camellia*

campanula'ria: bell-flowered; bell-shaped

campanula'tus: campanulate; bell-shaped →

campanuloi'des: resembling *Campanula*, the bell-flowers

campes'tris: growing in fields or plains; of the flatlands

camphora'tus: pertaining to or resembling camphor

MOUNTAIN SLIPPERS
Calceolatus. Cypripedium calceolatus is one of the rare and most beautiful wild orchids of the Blue Ridge Mountains. The plant takes its name from the inflated lip of the flower that forms a prominent pouch. In the South it is called the yellow lady's slipper.

✿

CAMELLIA FEVER
Camelliiflorus. "We in America did not get excited about camellias until the 1920s. After that time, camellia fever spread very fast, mainly because American men took the camellia as one of their favorite flowers. The cult spread across the South the way it had done on the West Coast."
William L. Hunt,
Southern Gardens, Southern Gardening, 1982

✿

STUDS FOR AN EMPIRE
Cappadocicum. Cappadocia was, in ancient times, a district of northeastern Asia Minor with shifting boundaries. Tiberius made it a Roman province in A.D. 17. It was thereafter imperially favored as the home of the emperors' stud racehorses.

HORN OF THE SOUTH ↑
Capricornis. The constellation Capricorn is visible only in the Southern Hemisphere sky and is the counterpart to Cancer. The Tropic of Capricorn is 23.5 degrees south of the equator, the southernmost point reached by the overhead sun.

campylocar'pus: with curved or bent fruit

camtschat'icus: of or from the Kamchatka Peninsula, Siberia

canaden'sis: of or from Canada or the northeastern United States

canalicula'tus: channeled; with a longitudinal groove

canarien'sis: of or from the Canary Islands

canar'ius: canary yellow

cancella'tus: crossbarred; like latticework; checkered

can'dicans: white; hoary; wooly

candidis'simus: very white or hoary

can'didus: pure white; shining

canes'cens: with very short hairs that give a whitish appearance

cani'nus: pertaining to a dog; sharp-toothed; indicating inferiority

cannab'inus: like *Cannabis* or hemp

cantab'ricus: of or from the Cantabrian Mountains, northern Spain

cantabrigien'sis: of or from Cambridge, England

ca'nus: ash-colored; off-white

capen'sis: of or from the Cape of Good Hope, South Africa

capillar'is: hair-like; very slender

capillifor'mis: shaped like hair

capil'lipes: with a slender stalk

capita'tus: forming a head

capitella'tus: having little heads

capitula'tus: having little heads

cappadoc'icum: of or from Cappadocia, eastern Asia Minor

capreola'tus: winding; twining; with or formed like tendrils

cap'reus: of the goat; smelling like the goat; fetid

capricor'nis: like a goat's horn; of or from or below the Tropic of Capricorn

cap'sicum: biting to the taste; hot (as peppers)

capsular'is: having capsules; enclosed by a membrane

caput-medu'sae: Medusa's head

cardaminefo'lius: with leaves like *Cardamine*; cress-like

cardia'cus: used for treating heart conditions

cardina'lis: cardinal red

cardiopet'alus: with heart-shaped petals

cardua'ceus: resembling the thistles →

cardun'culus: resembling a small thistle

caribae'us: of or from the Caribbean

carico'sus: like the sedges, *Carex*

carina'tus: keeled; ridged

carinif'era: keel-bearing

carmina'tus: carmine

car'neus: flesh-colored

car'nicus: fleshy

carnos'ulus: somewhat fleshy

carno'sus: fleshy; thick and soft

BITTER SALAD
Cardaminefolius. "Bitter Cress. *Cardamine clematis.* Of our several native and introduced species of Cardamine, this one has the largest flowers with petals ¼ inch or more long. Plants of some species, as the common name implies, are used for cress, or salad."
Justice and Bell, *Wild Flowers of North Carolina,* 1987

❧

WITHERED SEDGES
Caricosus. "The sedge has withered from the lake, ∾ And no birds sing."
John Keats, "La Belle Dame Sans Merci," 1820
Carex is the Latin name for a genus of perennial, grass-like herbs. For the amateur, sedges are distinguished from grasses by their smooth stems; grasses are jointed.

CARROT COIFFURE ↑
Carota. Daucus carota is
the wild carrot of the
roadside, known best
as Queen Anne's lace.
The cultivated form of
carrot is *D. carota* var.
sativa, as old as the
ancients. Parkinson
said fashionable ladies
wore carrot leaves in
their hair in place of
feathers in his time.
Nicholas Culpeper
recommended the
seeds, "being taken in
wine, or boiled in wine
and taken, it helpeth
conception."

carolinia'nus, caroli'nus: of or from the Carolinas,
southeastern United States

caro'ta: the Latin word for carrot

carpath'icus, carpat'icus: of or from the Carpathian
Mountains, eastern Europe

carpinifo'lius: with leaves like the hornbeam birch, *Carpinus*

cartilagin'eus: like cartilage in texture

caryophylla'ceus: with a clove-like fragrance; resembling the
pinks (*Dianthus, Silene,* and so on)

caryopteridifo'lius: with opposite, toothed leaves like vervain,
Caryopteris

caryotaefo'lius: with leaves like the fishtail palms, *Caryota*

cashmeria'nus: of or from Kashmir, south Asia

cas'picus, cas'pius: of or from the Caspian Sea

cassinoi'des: resembling *Cassine,* one of two shrubs: 1) *Ilex
cassine,* a North American holly; 2) *Cassinia,*
flowering shrub of the Southern Hemisphere

castane'us: chestnut brown

catalpifo'lius: with leaves like the *Catalpa*

catawbien'sis: of or from the Catawba River region of
North Carolina

cathar'ticus: cathartic; purging

cathaya'nus: of Cathay (China)

caucas'icus: of or from the Caucasus, the region between
the Black and Caspian seas

cauda'tus: tailed; with a tail-like appendage

caules'cens: having a distinct stem; having a stem aboveground

cauliala'tus: having a winged stem

cauliflo'rus: developing flowers directly on a stem

caus'ticus: burning to the taste

cayennen'sis: of or from Cayenne, French Guiana

celastri'nus: resembling the false bittersweet, *Celastrus*

cenis'ius: of Mount Cenis between France and Italy

centifo'lius: hundred-leaved; many-leaved

centranthifo'lius: with spurred leaves like Jupiter's-beard, *Centranthus*

cephala'tus: bearing heads

cephalon'icus: of Cephalonia (one of the Ionian Islands), Greece

cephalo'tes: with a small head-like appendage

cepifo'lius: with foliage like the onions

ceram'icus: ceramic; pottery-like; of or from the island of Ceram, Moluccas Islands, east Indonesia

cerasif'era: bearing cherries or cherry-like fruit

cerasifor'mis: formed like a cherry

cerasi'nus: cherry red

ceratocau'lis: with a horn-like stalk

cerea'le: pertaining to Ceres, goddess of agriculture

cerefo'lius: with waxy leaves

BLOOM ON THE BALD *Catawbiensis.* French botanist André Michaux (1746–1802) extensively explored the Piedmont and mountain regions of the Carolinas. He was the first to identify *Rhododendron catawbiense,* which comes into flower in early summer, setting mountain balds abloom.

FRUITLESS PURSUITS ↑ *Cerasiformis.* "It is no use making two bites of a cherry." Proverb

FLOWERS TO CARRY
Cheiri. Cheiranthus cheiri
are the wallflowers
that bloom through-
out the winter in the
southern states and
parts of England.
The Greeks picked
them for bouquets as
their name suggests—
cheiri for hand, *anthos*
for flower.

❦

SWALLOW BLOSSOMS
Chelidonioides. Chelido-
nium takes its name
from the Greek for a
swallow; the plants are
said to bloom at the
arrival of the first swal-
lows in spring.

❦

PROOF OF CHIMAERAS
Chimaera. The chimaera
was the fabled fire-
breathing monster of
the ancients, formed
variously of parts of
the goat, the lion, and
the serpent. Satan dis-
covered the chimaera
on his march across

ce'reus: waxy; wax-colored

cerif'era: wax-bearing

cer'inus: waxy; wax-colored

cer'nuus: drooping; nodding

cervi'nus: fawn-colored; tawny

chalcedon'icus: of or from Chalcedon on
    the Bosporus Strait opposite Istanbul

chalicodo'philus: growing in gravel

chatham'icus: of or from Chatham
    Island, New Zealand

cheilanthifo'lius: with leaves like the
    lip-fern, *Cheilanthes*

cheilan'thus: with lipped flowers

chei'ri: red-flowered

chelidonioi'des: like celandine, *Chelidonium* ↗

chersoph'ilus: growing in dry places

chi'a: of or from the island of Chios, now
    Khios, in the Aegean Sea

chilen'sis: of or from Chile

chiloen'sis: of or from Chiloe Island, off the coast of Chile

chimae'ra: fanciful; monstrous

chinen'sis: of or from China

chionan'thus: having snow white flowers; resembling the
    fringe tree, *Chionanthus*

chirophyl'lus: with hand-shaped leaves

chloran'thus: green-flowered

chlorochi'lon: green-lipped

chrysanthemoi'des: resembling the Chrysanthemums →

chrysan'thus: golden-flowered

chrys'eus: golden yellow

chrysocar'pus: golden-fruited

chrysoc'omus: golden-haired

chrysol'epis: golden-scaled

chrysoleu'cus: gold and white

chryso'lobus: golden-lobed

chrysophyl'lus: golden-leaved

chrysos'tomus: golden-mouthed

cibar'ius: edible; useful for food

cichoria'ceus: resembling chicory, *Cichorium*

cicutaefo'lius: with large, pinnately cut leaves like the water hemlock, *Cicuta*

cicuta'rius: of or like the water hemlock, *Cicuta*

cilia'ris, cilia'tus: fringed; with soft hairs

cilic'icus: of Cilicia, southern Turkey

ciliic'alyx: with a fringed calyx

cinc'tus: girded, girdled, or encircled

cinera'ceus: ash-colored; covered with gray hair

cinerariaefo'lius: with wooly leaves like *Cineraria*, now called *Senecio*

hell in *Paradise Lost* (1667): "Where all life dies, death lives and Nature breeds, ∾ Perverse, all monstrous, all prodigious things, ∾ Abominable, inutterable, and worse ∾ Than Fables yet have feign'd, or fear conceiv'd, ∾ Gorgons and Hydras, and Chimeras dire." Other writers of Milton's time had dismissed the possibility of such a creature and used the term for a self-deluding fancy. It has taken twentieth-century science to birth the monster anew. In 1926, *Popular Science* reported, "If the front half of one species be grafted on to the back half of another species, both continue to differentiate, and a chimaera or mosaic organism is produced." A chimaera at last is scientifically sound.

CINNAMON NESTS
Cinnamomifolius. "The process of collecting cinnamon is still more remarkable. . . . What they say is that the dry sticks, which we have learnt from the Phoenicians to call cinnamon, are brought by large birds, which carry them to their nests, made of mud, on mountain precipices . . . the method the Arabians have invented . . . is to cut up the bodies of dead oxen, or donkeys . . . which they . . . leave on the ground near the nests. They then retire to a safe distance and the birds fly down and carry off the joints of meat to their nests, which, not being strong enough to bear the weight, break and fall to the ground. Then the men come along and pick up the cinnamon."
Herodotus, *Histories*

cineras'cens: becoming ash gray

cine'reus: ash gray

cinnabari'nus: cinnabar red; vermilion

cinnamo'meus: cinnamon brown

cinnamomifo'lius: with leaves like the cinnamon tree, *Cinnamomum*

circina'lis, circana'tus: coiled; rolled down from the top like the new fronds of ferns

cirrha'tus, cirrho'sus: with large tendrils

cismonta'nus: on this (the south or Italian) side of the Alps

cistifo'lius: with leaves like the rockrose, *Cistus* ↓

citra'tus: citrus-like

citrifo'lius: citrus-leaved

citri'nus: citron-colored or citron-like

citriodo'rus: lemon-scented

citroi'des: citrus-like

clado'calyx: with a club-like calyx

clandesti'nus: concealed, hidden, or secret

clau'sus: shut; closed

clava'tus: club-shaped; thickened at the top

clavella'tus: slightly club-shaped

cla'vus: club

clematid'eus: with long climbing branches; like *Clematis*

clethroi'des: resembling the sweet-pepperbush, *Clethra*

clivo'rum: of the hills; growing on the slopes

clypea'tus: with or shaped like the round shield used
   by the Romans

clypeola'tus: slightly shield-shaped

cneo'rum: resembling a shrubby olive

coarcta'tus: crowded together; contracted

coccif'era, coccig'era: berry-bearing; hosting coccoid insects

coccin'eus: scarlet

cochenillif'era: yielding cochineal, a red dye

cochlea'ris: spoon-shaped

cochlea'tus: twisted or spiraled like a snail's shell

coelesti'nus: sky blue; heavenly

coeles'tis: sky blue; celestial

cogna'tus: closely related to

col'chicus: of or from Colchis, south of the Caucasians;
   poisonous

colea'tus: having a sheath

colli'nus: growing on a hill

colombi'nus: like a dove; flowers shaped like a group of doves

colo'nus: cultivated; forming a mound

coloraden'sis: of or from Colorado, United States

colora'tus: colored

columbia'nus: of or from British Columbia, Canada

columna'ris: columnar; pillar-like

co'mans, coma'tus: having hair; tufted with hair

commix'tus: mixed; mingled

WINDOW KISSING ↑
Clematideus. "Rose,
rose and clematis,
Trail and twine and
clasp and kiss."
Alfred, Lord Tennyson,
"Window," 1870
Tennyson's rhyme
found far more favor
with Elizabeth
Lawrence than another
pairing. "Occasionally
there is a choice [in
pronunciation]," Miss
Lawrence wrote in
Through the Garden Gate,
"between the correct
and the accepted, but
I have never found any
justification for Amy
Lowell's rhyming
clematis with window
lattice."

A RAGE FOR CONIFERS ↑
Conifera. The "conifer-
ization" of Great
Britain began with the
great plant explorers in
the late eighteenth cen-
tury. The rage to pos-
sess exotic trees was
spurred on when
Queen Victoria herself
planted a *Deodar* cedar,
the "divine tree" of
India. But by mid-
twentieth century,
horticulturists and
gardeners began to
doubt the wisdom of
such large-scale plant-
ing of imported
conifers.

commu'nis: growing in common; general

commuta'tus: changed or changing

como'sus: with long hair; hairy

compac'tus: compact; dense

complana'tus: compressed; flattened out on the ground

complex'us: encircled; embraced

complica'tus: folded

compos'itus: compound; made of two or more elements

compres'sus: compressed; flattened sideways as a stem

comp'tus: adorned; ornamented; with a headdress

con'cavus: hollowed out; basin-shaped

conchaefo'lius: with shell-shaped leaves

concin'nus: neat; well-made; of elegant appearance

con'color: uniformly colored; having matching color parts

condensa'tus, conden'sus: condensed; crowded

confertiflo'rus: with crowded flowers

confer'tus: crowded

confor'mis: similar in shape or otherwise; symmetrical

confu'sus: uncertain; easily mistaken for another species

conges'tus: congested; brought together

conglomera'tus: crowded together

congola'nus: of the Congo

conif'era: cone-bearing

conjuga'tus, conjugia'lis: connected; joined together in pairs

conna'tus: twin; with like parts united; united at the base *conoid'eus:* cone-like

conop'seus: cloudy; gnat-like

consanguin'eus: closely related

consol'idus: consolidated; stable

consper'sus: spattered; speckled

conspic'uus: easily seen; marked

constric'tus: drawn together; erect

contig'uus: together; touching

continenta'lis: continental; from the
   mainland

contor'tus: twisted

contrac'tus: contracted

controver'sus: controversial; disputed

convallarioi'des: resembling lily-of-the-valley,
   Convallaria →

convall'is: of the valley

convolvula'ceus: resembling the morning glory,
   Convolvulus

copalli'nus: gummy; resinous

coraci'nus: raven black

coralliflo'rus: with coral flowers

coral'linus: coral red

corda'tus: heart-shaped

cordifo'lius: with heart-shaped leaves

cordifor'mis: of a heart-like form

CULT OF THE LILY
Convallarioides. Colette
wrote of the lily-of-the-
valley in *For a Flower
Album:* "Its cult excites
the entire populace of
a capital city to a pitch
of effervescence. . . .
Come to Paris on May
Day and watch the
flower sellers' frontal
attack in the streets,
twenty francs the sprig,
a thousand francs the
bunch. . . . Their long
pale-green leaves are
always arranged as a
coronal round the
flowers; a tradition
no one dreams of
abolishing."

❦

SPANISH GUM
Copallinus. One of the
prizes of the Spanish
New World was a fra-
grant white resin called
copal that came from
various tropical trees,
mostly of the *Copaifera*
family.

❦

CORNISH PRAYERS
Cornubiensis. "'Tis heard where England's eastern glory shines ∽ And in the gulphs of her Cornubian mines." William Cowper, *Hope*, 1782

❧

CUCKOLD'S FLOWER
Cornutus. "He that thinks every man is his wife's suitor ∽ Defiles his bed, and proves his own cornutor." Jordan, *Poems*, 1675 Horns were the symbol of cuckoldry, and almost any horned flower could imply a husband deceived by his own wife.

❧

TO EACH HIS OWN
Croceus. Saffron crocus. *Crocus sativus.* "This is the flower that has been cultivated since most ancient times for the sake of the dried yellow stigmas, which were put to all manner

corea'nus: Korean

coria'ceus: leathery; tough and pliable; like leather

coria'ria: leather-like; used for tanning leather

cor'neus: horn-like

cornicula'tus: horned; having a horn-like structure

cornubien'sis: of or from Cornwall, southwest England

cornu'tus: horned

corolla'tus: corolla-like; resembling petals

corolli'nus: with a conspicuous corolla

coromandelia'nus: of Coromandel, India

corona'rius: used for or belonging to garlands or wreaths

corona'tus: crowned

corruga'tus: corrugated; wrinkled; furrowed

cor'sicus: of or from the island of Corsica

cortica'lis: with notable bark

cortico'sus: heavy with bark

corus'cans: vibrating; glittering

corylifo'lius: with leaves like the hazelnut, *Corylus* →

corymbif'era: with a corymb, a broad, flat flower cluster blooming from the outside

corymbo'sus: resembling a corymb

corynoc'alyx: with a club-like calyx

cosmophyl'lus: with leaves like *Cosmos*

costa'tus: ribbed; with a prominent ridge or vein

cotinifo'lius: with leaves like the smoke tree, *Cotinus*

coum: of the island of Cos, off the coast of Asia Minor

crassicau'lis: thick-stemmed

crassifo'lius: thick-leaved

cras'sipes: with a thick foot or stalk

crassius'culus: somewhat thick

cras'sus: thick; fleshy

crataegifo'lius: with leaves like the hawthorn, *Crataegus*

cre'brus: close, frequent, repeated, or crowded

crena'tus: scalloped; edged with rounded teeth

crenula'tus: a bit scalloped; having small, blunt teeth

crepida'tus: slipper-shaped; shoe-like

crep'itans: crackling; rustling

creta'ceus: inhabiting chalky soils;
    of or with the texture of chalk

cret'icus: of Crete in the eastern
    Mediterranean

crini'tus: with long hair

crispa'tus, cris'pus: crisped; curled; kinky

cristagal'li: cockscomb

crista'tus: crested; tasseled

croca'tus: saffron yellow

cro'ceus: saffron-colored; yellow →

crocosmaeflo'rus: with flowers like *Crocosmia*

crotonifo'lius: with brightly colored leaves like *Croton*

of domestic uses, medicinal, culinary, as a dye and as a perfume. . . . 'To the nations of Eastern Asia, its yellow dye was the perfection of beauty, and its odour a perfect ambrosia,' [*A Modern Herbal,* Mrs. M. Grieve]. . . . When I went to South America, many of the dishes served us were flavored with Saffron, and very unpleasant I thought it. A dingy neighborhood of London still is known as Saffron Hill, and the town of Saffron Walden, in Essex, commemorates the . . . Saffron industry. And so the Saffron Crocus is one of the most famous of flowers: but it is sadly disappointing in the garden."
Louise Beebe Wilder, *Adventures with Hardy Bulbs,* 1989

A PRETTY PICKLE
Cucumerinus. "Why
cucumber sandwiches?
Why such reckless
extravagance in one
so young?"
Oscar Wilde,
*The Importance of Being
Earnest*, 1895

❦

LIFE EVERLASTING
Cupressinus. "'Tis one
thing for a soldier to
gather laurels—and
'tis another to scatter
cypress."
Laurence Sterne, *The
Life and Opinions of Tris-
tam Shandy*, 1759–1767
Cypress wood is
almost everlasting—
the great cypress gates
of Constantinople
lasted 1,100 years.
Thus it was associated
with the life everlasting
after death. Hearses
were wreathed in
cypress and coffins
made from the wood.

❦

crucia'tus: in the form of a cross

crucif'era: bearing or marked with a cross; like the crucifers

cruen'tus: bloody; stained blood red in color

crusgal'li: cockspur

crusta'tus: encrusted; forming a hard and brittle crust

cryptan'drus: with hidden stamens

crystal'linus: crystalline; with a glistening surface

ctenoi'des: resembling a comb

cuculla'tus: covered with a hood

cucumeri'nus: resembling a cucumber

cucurbiti'nus: resembling a melon or gourd

culto'rum: of the cultivators or gardeners

cultra'tus: shaped like a curved
   knife blade

cunea'tus: wedge-shaped

cuneifo'lius: with wedge-shaped leaves

cuneifor'mis: with a wedge-like form

cuprea'tus: with the color of copper

cupressifor'mis: with the form of cypress

cupres'sinus: cypress-like

cupressoi'des: cypress-like

cu'preus: copper-like or copper-colored

cupula'ris: cup-shaped

curassav'icus: of Curaçao, West Indies

cur'tus: abruptly shortened; broken off

curva'tus: curved; arched; bent

curvifo'lius: with curved leaves

cuscutaefor'mis: resembling dodder, *Cuscuta* ↘

cuspida'tus: with a cusp or sharp, rigid point

cuspidifo'lius: leaves with a sharp, rigid point

cyanan'thus: with dark blue flowers

cya'neus: dark blue

cyanocar'pus: bearing blue fruits

cyanophyl'lus: with blue leaves

cyatheoi'des: like the tree ferns, *Cyathea*

cyathoph'ora: cup-bearing

cyclamin'eus: resembling the *Cyclamen;* coiled
   like the fruit stalk of *Cyclamen*

cyclocar'pus: with fruit arranged in a circle

cy'clops: cyclopean; gigantic

cylindra'ceus, cyclin'dricus: cylindrical

cymbifor'mis: boat-shaped; having a hollow or
   shallow recess

cymo'sus: bearing cymes, more or less flattened
   flower heads blooming from the middle out

cynaroi'des: resembling the thistle and artichoke,
   Cynara; bitter to the taste

cy'preus: copper-like

cyp'rius: of or from the island of Cyprus

cytisoi'des: resembling the brooms, *Cytisus*

MEASURING THE SKY
Cyaneus. A cyanometer,
the invention of Mon-
sieur Saussure, mea-
sures the intensity of
the blue of the sky.

❦

GNOMES IN CONCLAVE
Cyclamineus. "Narcissus
cyclamineus. A little horticultural joke . . .
thrusting forward its
abnormally long and
narrow tube with the
perianth turned
straight back in a man-
ner that has been vari-
ously compared to the
ears of an angry mule,
a frightened rabbit, a
kicking horse. And
indeed it does have a
startled expression. A
group of them makes
one think of a bevy of
gnomes in agitated
conclave."
Louise Beebe Wilder,
*Adventures with Hardy
Bulbs,* 1989

❦

PLUMS AND ROSES
Damascenus. "Upon
her head a Cremosin
coronet, ∾ With
Damaske roses and
Daffadillies set: ∾
Bayleaves betweene, ∾
And Primroses greene
∾ Embellish the
sweete Violet. . . . ∾
And if you come
hether, ∾ When
Damsines I gether, ∾
I will part them all
you among."
Edmund Spenser,
"The Lay to Elisa,"
1591
Both the damask rose,
Rosa damascena, and the
damson plum, *Prunus
domestica* var. *institia,* take
their English names
from Damascus, whose
fruits and flowers have
been celebrated since
ancient times. The
damask rose is the
source of the coveted
attar of rose.

❧

acrydioi'des: resembling the pine-like trees of
New Zealand, *Dacrydium;* yielding resin

dactylif'era: having finger-like appendages

dactyloi'des: resembling fingers

dahu'ricus, dau'ricus, davu'ricus: of or from Dauria, near
the Siberian-Mongolian border

dalmat'icus: of or from Dalmatia

damasce'nus: of Damascus, Syria;
with the color of *Rosa damascena*

danaeifo'lius: with leaves like the
Alexandrian or poet's laurel, *Danae*

dan'icus: of or from Denmark

daphnoi'des: resembling the fragrant
shrub, *Daphne →*

dasyacan'thus: with thick spines

dasycar'pus: with thick fruits

dasy'clados: with thick branches

dasyphyl'lus: with thick leaves; with hairy leaves

dasyste'mon: with thick stamens

daucoi'des: resembling the carrot, *Daucus*

dealba'tus: whitewashed; covered with a white powder

deb'ilis: weak, frail, or feeble

decalva'tus: made or becoming smooth-skinned; glabrous

decan'drus: with ten stamens

decapet'alus: with ten petals

decaphyl'lus: with ten leaves

decid'uus: shedding leaves annually; not evergreen

decip'iens: deceptive

declina'tus: bent downward from the base; turned aside

decolor'ans: discoloring; staining; losing color

decompos'itus: divided more than once

dec'orans: adorning, decorative, or handsome

decora'tus: decorative

decor'ticans: shedding bark

deco'rus: elegant, comely, or becoming

decum'bens: trailing on the ground with tips turned up

decur'rens: running down the stem, as a leaf whose base
　　runs down the stem beyond the point of insertion

deflex'us: bent abruptly downward, backward, or outward

defor'mis: misshapen; deformed

dehis'cens: splitting open to scatter seeds or pollen

dejec'tus: debased; fallen

delec'tus: chosen; choice

delicatis'simus: very delicate; most charming

delica'tus: delicate; tender

delicio'sus: delicious; of fine flavor; offering great pleasure

del'phicus: of or from Delphi, Greece

delphinifo'lius: with leaves like *Delphinium*

DOLPHIN FLOWER ↓
Delphinifolius. Delphinium
is a diminutive of the
Greek for dolphin
because the flower
nectary is said to
resemble the friend
of shipwrecked sailors.
The leaves of the plant
are less friendly,
known to poison
man and beast alike.

BURIED TEETH

Dens-canis. Erythronium dens-canis. "The Dog's-Tooth Violet is a misnamed plant, for the root, which resembles a dog's tooth, is invisible above ground, while the flower has not the remotest likeness to a violet, except that it blooms in the spring. In the old days it was associated with certain orchids because of its spotted leaves, and called, like them, satyrions (after the satyrs, because of some supposed aphrodisiac properties), a name that seems to suit it better, at least as regards the European species; the American ones have a somewhat more innocent look." Alice M. Coats, *The Treasury of Flowers,* 1975

deltoi'des, deltoid'eus: triangular, like the fourth letter of the Greek alphabet, *delta*

demer'sus: living underwater; submerged

demis'sus: low; weak; hanging down

dendric'olus: dwelling in the trees; epiphytic

dendroid'eus: tree-like

dendro'philus: growing on or around trees

densa'tus: dense; crowded

dens-can'is: shaped or like a dog's tooth →

densiflo'rus: densely flowered

densifo'lius: densely leaved

den'sus: dense; crowded

denta'tus: toothed, usually with sharp, outward pointing teeth

denticula'tus: slightly toothed

dentif'era: tooth-bearing

dento'sus: toothed

denuda'tus: naked; worn away

deodar'us: of or pertaining to the deodar, *Cedrus deodara*

deo'rum: of the gods

depaupera'tus: starved; dwarfed; imperfectly developed due to conditions

depen'dens: hanging down

depres'sus: flattened out from above

derelic'tus: abandoned; neglected

deser'ti: of the desert

desmoncoi'des: resembling the spiny feather palms, *Desmoncus*

deton'sus: clipped; shaved; bald

deus'tus: burned; scorched

diabol'icus: of the devil; having two-horned fruit

diacan'thus: with two spines or thorns

diade'ma: possessing a diadem or crown; crown-like

dian'drus: two-stamened

dianthiflo'rus: with flowers like the pinks, *Dianthus* →

diaph'anus: transparent

dichot'omus: forked in pairs; repeatedly divided into two equal parts

dichroan'thus: with flowers like *Dichroa;* with flowers of two distinct colors

dichro'us: of two distinct colors

dicoc'cus: furnished with two berries or nuts

dictyophyl'lus: with netted leaves; exhibiting a conspicuous network of veins

did'ymus: in pairs; twinned

diffor'mis: of differing forms; of unusual shape

diffu'sus: loosely spreading; branching from the axis at an angle between 45 and 90 degrees

digita'tus: hand-like; fingered; shaped like an open hand

A DEVIL'S GARDEN
Diabolicus. Devils have invaded the garden since Eden, but some of these must be welcome for so many plants have taken devil for their common names. A devil's garden: devil flower, *Tacca chantrieri;* devil's backbone, *Kalanchoe daigremontiana;* devil's claw, *Physoplexis comosa;* devil's fig, *Argemone mexicana;* etc., all set perhaps within serpentine walls?

❧

OR AN ANGEL'S?
Dianthiflorus. The genus name *Dianthus* means flower of the gods, and if the idea of a devil's garden isn't appealing, perhaps a heavenly one is. Beyond the many *Dianthus,* there's an angelic host of plants with heaven-inspired nomenclature.

❧

A CUP FOR SHEPHERDS ↑
Dipsaceus. Teasel, *Dip-sacus follonum.* A useful plant, following its names. *Dipsa* means thirst in Greek and was given to this plant, which collects and holds water in the leaf base, a cup for shepherds. *Follonum* is from the Latin for fuller, those who stretch and tease out wool for spinning. The fullers in England used the spiny, dried seed heads of teasel as combs.

dilata'tus, dila'tus: expanded; spread out; widened

dilu'tus: washed out; thinned out (as of color); diluted

dimidia'tus: lopsided; halved with only one side developed

dimor'phus: existing in two forms; two forms of the same structure (as a leaf) occurring on the same plant

di'odon: with two teeth

dioi'cus: having separate male and female plants

diosmaefo'lius: with leaves like the fragrant shrub, *Diosma*

dipet'alus: having two petals

diphyl'lus: having two leaves

dipsa'ceus: resembling teasel, *Dipsacus*

dipterocar'pus: producing two-winged fruit

dip'terus: having two wings or wing-like appendages

dipyre'nus: having fruit with two stones or kernels

discoi'deus: without rays; disc-like

dis'color: of two or more different colors; variegated

dis'par: dissimilar; with unexpected characteristics

dissec'tus: deeply cut; divided into deep lobes or segments

dissim'ilis: unlike the expected characteristics of the genus

dissitiflo'rus: with loose flower heads

distach'yus: having two spikes or branches

dis'tans: widely separated; remote

distichophyl'lus: with leaves arranged in two ranks

dis'tichus: with flowers or leaves in two opposite rows

dis'tylus: having two styles

diur'nus: day-flowering; with flowers lasting one day; ephemeral

divarica'tus: spreading at a wide angle; straggling

diver'gens: going different ways

diversic'olor: diversely colored

diversiflo'rus: having flowers of two or more forms

diversifo'lius: having leaves of two or more forms

divionen'sis: of or from Dijon, France

divi'sus: divided

di'vus: belonging to the gods

dodecan'drus: with twelve stamens

dolabra'tus: hatchet-shaped

dolabrifor'mis: hatchet-shaped

doliiform'is: shaped like a barrel

dolo'sus: deceitful; appearing to be another plant

domes'ticus: domesticated; for use as a houseplant

doronicoi'des: resembling leopard's-bane, *Doronicum*

drabifo'lius: with leaves like the crucifer, *Draba*

dracaenoi'des: resembling the ornamental variegated *Dracaena*

dracoceph'alus: shaped like or resembling a dragon's head

dracunculoi'des: resembling tarragon, *Dracuncula*; dragon-like

drepanophyl'lus: with sickle-shaped leaves

drupa'ceus: developing fruits with stones, as the peach or cherry

drupif'era: bearing a drupe

CIVILIZED ROOTS
Doronicoides. "Civilization—spurns—the Leopard! ∾ Was the Leopard—bold?" Emily Dickinson, "Civilization," 1862 *Doronicum* is called leopard's-bane from its use as a poison to bothersome wild animals. The roots were a staple in the apothecary's shop.

DRAGON'S GARDEN ↑ *Dracocephalus.* There are as many dragons as devils in the garden. Beyond the snapping ones are all whose names come from *drakon,* Greek for dragon.

drynarioi'des: resembling the oakleaf fern, *Drynaria*

dryophyl'lus: with leaves like the oaks ↘

du'bius: doubtful; uncertain; nonconforming

dulcama'ra: bittersweet to the taste

dul'cis: sweet to the taste; mild

duma'lis: thorny; bushy; shrubby

dumeto'rum: of bushes or hedges

dumo'sus: bushy

dunen'sis: growing in the sand dunes

du'plex: double; growing in pairs

duplica'tus: duplicate; double

durab'ilis: durable; lasting

durac'inus: producing hard fruits or berries

durius'culus: somewhat hard or rough

du'rus: hard

dysenter'icus: indicating plants used to treat dysentery; reputed to cause dysentery

ORACULAR OAKS ↑
Dryophyllus. The Dryads inhabited the oak trees from which they took their name, *drys* being the Greek for oak. They were the daughters of Zeus to whom the oak was sacred. Most sacred was the grove at Dodona, the most ancient of all oracles, where the will of the god was interpreted through the rustlings of the leaves.

bena'ceus: resembling ebony, the name for the dark inner wood of several tropical trees

ebe'nus: ebony black

eboracen'sis: of or from York, England, formerly Eboracum

ebractea'tus: without bracts, leaflets at the base of flowers

ebur'neus: ivory white

echina'tus: prickly; bristly like a hedgehog, *echinos* in Greek

echinocar'pus: bearing prickly fruits

echinosep'alus: producing prickly sepals

echioi'des: resembling borage, a long-leaved herb with blue flowers, or viper's-bugloss, *Echium* ↓

ecornu'tus: without horns

edenta'tus: without teeth

edu'lis: edible; of food

effu'sus: very loose-spreading; straggling in growth

elaeagnifo'lius: with leaves like the Russian olive, *Elaeagnus*

elas'ticus: returning to the original position when pulled or bent; yielding an elastic substance

elata'rius: driving out with force; propelling seeds

ela'tus: tall

elec'tus: picked out; selected

el'egans: elegant

NIGHTMARE DARK

Ebenaceus. Ebony is the wood for the furniture and fixtures of the night, according to the poets. Night sits on "her ebon throne" and Fate's "ebon sceptre rules The Stygian deserts." And in Shelley's "ebon Mirror, nightmare fell."

❦

HEDGEHOG BUSHES

Echinatus. "I remember planting echinops because I like to say with the poet Vernede, 'It is July in my garden and steel-blue are the globe thistles'.... Their spiky form gives the plant its name—from the Greek—which means 'resembling a hedge-hog.'"
Elizabeth Lawrence, *Through the Garden Gate,* 1990

❦

HEATHLAND FERNS ↑
Ericetorum. "Now, when
alone, do my thoughts
no longer hover ∾
Over the mountains,
on that northern
shore, ∾ Resting their
wings where heath and
fern-leaves cover ∾
Thy noble heart for
ever, ever more?"
Emily Brontë,
"Remembrance," 1846

❧

elegantis'simus: most elegant

elegan'tulus: elegant

elephant'idens: with large teeth

elephan'tipes: elephant-footed

elephan'tum: of the elephants; monstrous in size

ellipsoida'lis: elliptical

ellip'ticus: elliptical; about twice as long as wide

elo'des: of bogs and marshes

elonga'tus: elongated; lengthened

emargina'tus: notched at the apex as though a small piece
had been removed, said of leaves

emet'icus: causing vomiting

em'inens: eminent; prominent; with a noticeable projection

emoden'sis: of or from Mount Emodus, western Himalayas

endi'via: Latin for chicory

enerv'is: apparently lacking veins

enneacan'thus: nine-spined

enneaphyl'lus: nine-leaved

ensa'tus: sword-shaped; two-edged and tapering

ensifo'lius: sword-leaved

entomoph'ilus: pollinated by insects; attracting insects by
color or fragrance

epi'gaeus: growing close to the ground; on dry land

epihy'drus: growing on the water's surface

epilith'icus: growing upon the surface of stones

epiphy'ticus: growing upon another plant ↓

epite'ius: annual; yearly

epixy'lous: growing upon wood

eques'tris: pertaining to the horse

equi'nus: of horses

equisetifo'lius: with leaves like the horsetails, *Equisetum*

erec'tus: erect; upright

eriacan'thus: with wooly spines

erianthe'ra: with wooly anthers

erian'thus: with wooly flowers

ericaefo'lius, ericifo'lius: with leaves like the evergreen heaths, *Erica*

ericeto'rum: of the heathlands

ericoi'des: *Erica*-like; heath-like

erina'ceus: like a hedgehog; prickly

eriobotryoi'des: resembling the loquat, *Eriobotrya*

eriocar'pus: wooly-fruited

erioceph'alus: wooly-headed

erioph'orus: bearing wool

eriospa'thus: wooly-spathed

eriostach'yus: with a wooly spike

erioste'mon: with wooly stamens

eristha'les: very luxuriant

ero'sus: jagged, as if gnawed

MISSING LINKS

Equisetifolius. Equisetums, or horsetails, are à sort of missing link in the plant world. In the Carboniferous period, extensive forests of gigantic woody *Equisetum*-like plants flourished. Their fossilized remains form most of the modern coal beds. The modern horsetail is a rush-like plant with rough stems growing near rivers and streams. They once were gathered for household scrubbing and are used in cabinet work in place of sandpaper.

❧

THE BURNING ROOT

Ericoides. Erica arborea is a Mediterranean heath whose roots are dug for briar, which has nothing to do with thorns, but is a corruption of *bruyère,* French, to burn. The wood is used to make tobacco pipes.

❧

FOR BROKEN BONES ↑
Eupatorioides. The *Eupatoriums* provide a great late-summer show throughout the Southern states. *Eupatorium fistulosum*, joe-pye weed, reaches seven feet or more with flower heads more than a foot across. Bill Hunt writes that the native Americans boiled the roots of joe-pye and washed newborn infants in the water to give them strength and vigor matching that of the plant. *E. perfoliatum* was called boneset for its use as a treatment for broken bones.

errat'icus: erratic; unusual; sporadic; of no fixed habitat

errome'nus: vigorous in growth; robust

erubes'cens: blushing; turning red

erucoi'des: resembling the rocket-green, *Eruca*

erythrae'us: red

erythrocar'pus: red-fruited

erythroceph'alus: red-headed

erythrop'odus: red-footed; red-stalked

erythrop'terus: red-winged

esculen'tus: edible; tasty

estria'tus: without stripes

etrus'cus: Etruscan; of Tuscany, the classical Etruria

etubero'sus: lacking tubers

euboe'us: of or from the Greek island of Euboea

eucalyptoi'des: resembling *Eucalyptus*

euchlor'us: of a beautiful green color

euchro'mus: well-colored

eugenioi'des: resembling the tropical clove tree, *Eugenia*

euo'des: sweet-scented

eupatorioi'des: resembling mistflower, *Eupatorium*

euphle'bius: with prominent or numerous veins

euphorbioi'des: resembling the spurges, *Euphorbia*

eu'podus: having a long stalk

europae'us: European

eustach'yus: with long tresses of hair

evanes'cens: disappearing quickly

evec'tus: extended

ever'tus: expelled abruptly; turned out; turned inside out

exalta'tus: exalted; very tall

exara'tus: furrowed; with grooves; appearing to be engraved

excava'tus: excavated; hollowed out

excel'lens: excellent; distinguished

excel'sior: taller

excel'sus: tall, lofty, or high

exci'sus: cut away; cut out

excortica'tus: stripped of bark

exig'uus: little, poor in growth, or weak

exim'ius: distinguished; out of the ordinary; excellent

exitio'sus: pernicious; destructive; fatal; deadly

exole'tus: mature; dying away without contents, as a fruit

exonien'sis: from Exeter in Devon, England

exot'icus: exotic; from another country

expan'sus: expanded; spread out

explo'dens: exploding

exsca'pus: without a scape

exsculp'tus: dug out; with deep cavities; chiseled

exser'tus: protruding from or beyond surrounding organs

exsur'gens: rising; rising out of

exten'sus: extended; wide

exu'dans: producing a sticky secretion

WART-EATERS ↑
Euphorbioides. Pliny wrote "Iuba, king of Mauritania, found out this herb *Euphorbia*, which he so called after the name of his own Physitian Euphorbus." Culpeper says "Spurges are mercurial plants, and abound with a hot and acrid juice, which, when applied outwardly, eats away warts and other excrescences." A gum resin, *Euphorbium*, still used medicinally as an emetic and purgative is made from *E. offici-narum, E. antiquorum,* and *E. canariensis.*

BEAN SALAD
Fabaceus. "Salad of Broad Beans [Fava beans], Cucumber, and Oranges—½ lb. small new raw broad beans, ½ a cucumber, 2 oranges, a few radishes. Shell the beans, cut the unpeeled cucumber into small squares, the oranges into quarters, and the radishes into thin rounds. Mix them with a vinaigrette sauce made with an egg. Good with cold duck." Elizabeth David, *Summer Cooking*, 1955

aba'ceus: resembling the broad bean called faba, *Vicia faba*

face'tus: elegant; fine; humorous

fagifol'ius: with leaves resembling the beech, *Fagus* ✓

falca'tus: shaped like a sickle; curved

falcifo'lius: with sickle-shaped leaves

fal'lax: deceptive; false

farca'tus: stuffed or filled; not hollow; like a soft pod

farina'ceus: yielding starch; of a mealy texture

farino'sus: with a mealy or powdery surface

fascia'tus: having parts that have grown together and flattened out, as several stems; bundled together; marked with broad bands of color

fascicula'tus: clustered; grouped together

fascina'tor: fascinating; bewitching

fastigia'tus: with erect branches growing close together; columnar; upright

fastuo'sus: proud; haughty

fat'uus: foolish; insipid; worthless

favo'sus: honeycombed; with regular surface cavities

febrif'ugus: having the ability to alleviate fevers

fecun'dus: fertile; fruitful

fejeen'sis: of or from the Fiji Islands, South Pacific

felos'mus: foul-smelling

fem'ina: female

fenestra'lis: pierced by window-like openings

fenn'icus: of or from Finland, formerly Fennica

fe'rax: fruitful

fe'rox: ferocious; very thorny

fer'reus: pertaining to iron; durable; hard
as iron

ferrugin'eus: rust-colored

fer'tilis: fertile, fruitful, or heavy-seeding

ferula'ceus: resembling the giant fennel,
Ferula; with hollow stems

fe'rus: wild; reverting to the wild state

festi'vus: festive, gay, or bright

feti'dus: stinking

fibrillo'sus: composed of fibers; fibrous

fibro'sus: having prominent fibers

ficifo'lius: with leaves resembling the fig, usually *Ficus carica*

ficoi'des, ficoid'eus: resembling the fig, usually *Ficus carica* ↑

filamento'sus: thread-like

filicau'lis: thread-stemmed

filici'nus: resembling the ferns, formerly broadly grouped
as the family *Filices*

filicoi'des: resembling ferns

filif'era: composed of or bearing thread-like structures

HELP FOR CHILBLAINS
Ficoides. Culpeper thought figs "fitter for medicine than for any other profit that is gotten by the fruit of them. . . . A decoction of the leaves being drunk inwardly, or rather a syrup made of them, dissolves congealed blood caused by bruises or falls," mixing the ashes with lard "helps kibes and chilblains." Poets thought less of the fig. Pistol, in Shakespeare's *The Life of King Henry V* says, "Die and be damn'd! and figo for thy friendship!" To "make a fig" was a gesture of contempt. A "Spanish fig" was a poisoned one: "This boy . . . long he shall not soe, if figs of Spain . . . their force retain." (*The Time's Whistle,* 1616) And Cleopatra's asp always arrives in a basket of figs in *Antony and Cleopatra.*

Florentinus. Iris florentina is the source of orris, the powdered root still used as a fixative in perfumes and cosmetics. The rhizome, when dried, has the fragrance of violets; in modern perfumery, almost all "violet" scents are derived from orris.

❧

ELEMENT OF MYSTERY
Floridus. Calycanthus floridus. "All flower perfumes of great distinction . . . are from blossoms of modest color and bearing. The *Calycanthus*, called Virginia Allspice, Sweet Shrub, or Strawberry bush . . . has an element of mystery in it—that indescribable quality felt by children, and remembered by prosaic grown folk." Alice Morse Earle, *Old Time Gardens*, 1901

❧

filifo'lius: with thread-like leaves

filipenduli'nus: suspended by a thread; resembling *Filipendula* whose tubers are attached by slender filaments ↓

fimbria'tus: fringed

fimecar'ius: growing on dung

firma'tus: firm; made firm

fir'mus: firm; strong; steady

fissifo'lius: with split or cleft leaves

fis'silis: cleft or split

fis'sus: cleft or split almost to the base, as a leaf

fistulo'sus: tubular; pipe-like

flabella'tus: shaped like an open fan

flabel'lifer: bearing fan-shaped structures

flac'cidus: soft, limp, or feeble; collapsing under its own weight

flac'cus: drooping

flagella'ris, flagella'tus: whip-like; with long, thin shoots

flagel'lum: shaped like a whip

flam'meus: flame red in color

flaves'cens: yellowish; pale yellow

flav'idus: yellow; yellowish

flavispi'nus: with yellow spines

flavis'simus: deep yellow; very yellow

flavovi'rens: greenish yellow

fla'vus: pure yellow

flexicau'lis: with pliant or bent stems

flex'ilis: flexible, pliant, or limber

flexuo'sus: winding; growing in a zigzag manner

flocco'sus: wooly; looking like matted wool

flo'ra: in honor of Flora, Roman goddess of
flowering plants

flo're-al'bo: with white flowers

florenti'nus: of or from Florence, Italy

flo're-ple'no: with double flowers; full-flowered

floribun'dus: free-flowering; abounding in flowers; flowering
for a long season

florida'nus: of or from Florida, United States

flor'idus: flowering; full of flowers

flos-cu'culi: flowering when the cuckoo sings

flos-jovis: Jove's flower

flu'itans: floating on water; growing on water's surface

fluminen'sis: of or from a river; growing in running water;
of or from Rio de Janeiro, *flumen Januarii* in Latin

fluvia'tilis: of or from a river; growing in running water

foem'ina: feminine

foenicula'tus: resembling the culinary fennel, *Foeniculum*

foe'tidus: bad-smelling; stinking

folia'ceus: resembling a leaf; with many leaves

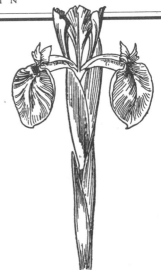

ROAST BEEF PLANT ↑
Foetidus. Iris foetidissima,
Theophrastus's stink-
ing iris, was used
medicinally by 300 B.C.
Modern medicine pre-
scribes an infusion of
the rhizomes for anti-
spasmodic and cathar-
tic conditions.
Culpeper thought the
scent of the bruised
leaves "evil," but oth-
ers detected the odor
of roast beef and
called it the roast beef
plant.

❦

FOR HER BREAKFAST *Fragarioides.* "The work in the vegetables—Gertrude Stein was undertaking the care of the flowers and box hedges—was a full-time job and more. Later it became a joke, Gertrude Stein asking me what I saw when I closed my eyes, and I answered, Weeds. That, she said, was not the answer, and so the weeds were changed to strawberries. The small strawberries, called by the French 'wood strawberries,' are not wild but cultivated. It took me an hour to gather a small basket for Gertrude Stein's breakfast . . . our young guests were told that if they cared to eat them they should do the picking themselves." Alice B. Toklas, *The Alice B. Toklas Cookbook,* 1954

folia'tus, folio'sus: leafy; full of leaves

follicula'ris: producing follicles, dry, one-chambered seed capsules that split on one edge

fontina'lis: growing by or near a spring of water

forfica'tus: shaped like scissors; deeply notched

formicar'ius: pertaining to ants; attracting ants

formosa'nus: of or from Taiwan, formerly Formosa

formosis'simus: very beautiful

formo'sus: beautiful; handsome; finely formed

for'tis: strong; growing vigorously

fourcroy'des: resembling *Furcraea,* an *Agave*-like plant

fovea'tus: pitted on the surface

fragarioi'des: resembling the strawberry, *Fragaria* →

frag'ilis: fragile; brittle

fra'grans: scented; especially sweet-scented

fragrantis'simus: very fragrant

fraxin'eus: resembling the ash tree, *Fraxinus*

frig'idus: growing in the cold regions

fructif'era: fruit-bearing; fruitful

fructig'enus: fruitful

frumenta'ceus: producing grain

frutes'cens, fru'ticans: shrubby; shrub-like

fuca'tus: painted; dyed

fuchsioi'des: resembling *Fuchsia* ⟍

fu'gax: fleeting; wasting away quickly; withering

ful'gens: shining or glistening; often, with red flowers

fulg'idus: shining or glistening

fuligino'sus: sooty; dirty brown to black in color

fullo'num: pertaining to fullers, persons who clean, shrink, and thicken cloth

ful'vidus: slightly tawny

ful'vus: tawny; brownish orange to a light reddish brown

fumarioi'des: resembling plants of the fumitory family, *Fumaria*

fumi'dus: smoky gray

funal'is: twisted together; rope-like

fu'nebris: funereal; of or from graveyards

fungo'sus: spongy; fungus-like

funicula'tus: like a slender rope or cord

fur'cans, furca'tus: forked

furfura'ceus: scurfy, flaking, or scaly

fur'iens: exciting to madness

fus'cus: brown; dusky; swarthy

fusifor'mis: shaped like a spindle, thickest in the middle and drawn out toward the ends

futil'is: useless

EARTH SMOKE
Fumarioides. Fumitory. Earth Smoke. *Fumaria officinalis.* The common name of this herb refers to the odor of the leaves and/or the color of its foliage. In ancient Britain, the plant was burned, its smoke dispelling witches and their evil spells. The Shakers much marketed the herb in this country in the nineteenth century, recommending it for "jaundice, obstruction of the bowels, scurvy, and in general debility of the digestive organs."
Amy Miller, *Shaker Herbs,* 1976

CHRISTMAS GALAXIES
Galacifolius. "Alice
Lounsberry says (in
*Southern Wild Flowers and
Trees*) Mrs. Kibbee, the
widow of a doctor
who died in a yellow
fever epidemic, was the
first person to think
of selling the leaves
[of *Galax*] at Christ-
mas, and that she
largely supported her
children by the sales.
In May, when the
creamy flowers stand
on thin stems above
the glossy new leaves,
Mrs. Lounsberry says
they look like a milky
way. I suppose she
thought of this
because the generic
name comes from the
Greek word for milk,
or because the moun-
tain people call it
galaxy."
Elizabeth Lawrence,
Through the Garden Gate,
1990

Gadita'nus: of or from Cádiz, Spain

galacifo'lius: with leaves like *Galax*

galan'thus: milk flower (white-colored)

gal'binus: yellowish green

galea'tus: having a helmet-like covering;
helmet-shaped

galegifo'lius: with leaves like the goat's-rue, *Galega* ↓

galioi'des: resembling bedstraw, *Galium*

gal'licus: of or from Gaul or France; also pertaining to a
cock or a rooster

ganget'icus: of or from the Ganges or the Gangetic Plain in
India and Bangladesh

gargan'icus: of or from Gargano, Italy

gel'idus: growing in the icy cold;
from cold regions

gemina'tus: twin; united in pairs

geminiflo'rus: having flowers arranged
in pairs

gemma'tus: bearing buds, especially
buds that are capable of
reproducing the plant; jeweled

gemmif'era: bud-bearing

genera'lis: general, normal, or as expected

geneven'sis: of or from Geneva, Switzerland

genicula'tus: jointed; bent like a knee at a node

genistifo'lius: with leaves like the broom, *Genista*

geocar'pus: with fruit ripening in the earth

geoi'des: of the earth; resembling *Geum*

geomet'ricus: marked with a regular or formal pattern

geophi'lus: ground-loving; spreading horizontally

georgia'nus: of the Caucasus, Georgia

geranioi'des: resembling the true *Geranium*

german'icus: German ↘

gibbero'sus: humped on one side; hunchbacked

gibbo'sus, gib'bus: swollen on one side;
    with a pouch-like swelling

gibraltar'icus: of or from Gibraltar

gigante'us: gigantic; very large

gigan'thes: with huge flowers

gi'gas: of the giants; immense

gla'ber: smooth; without hair, scales,
    or bristles

glaber'rimus: very smooth; completely
    smooth

glabra'tus: somewhat glabrous

glacia'lis: from the glacial regions,
    especially if found near glaciers

gladia'tus: like a sword

BATTLE BROOMS
Genistifolius. Both *Genista* and *Cytisus* are the brooms and they are scarcely distinguished by the amateur. Classification depends on a small protuberance on the seed. Both like exposure, growing in full sun and wind. One or the other—both formerly called *Planta Genista*—became the emblem of the Plantagenets who wore it in their hats in battle. The Scotch broom is *Cytisus scoparius;* Spanish broom is sometimes *Genista hispanica,* but more often, Spanish broom or weaver's broom is *Spartium junceum* whose coarse fibers can be used in clothmaking.

Glaucus. "Southern gardeners seem to have forgotten how good plums are in the summer months . . . served as dessert in Europe, they are so beautiful that one hardly wants to eat them. . . . 'Reine Claude,' 'Golden Drop,' 'Victoria,' 'Belgian Purple'—each is a work of art. They look as if they had been painted with watercolors. The Europeans are so appreciative of the beauty of these fruits that they pick them by the stem in order not to disturb the beautiful bloom that covers each fruit." William L. Hunt, *Southern Gardens, Southern Gardening,* 1982

glandifor'mis: in the form of a gland

glandulif'era: gland-bearing; glandular

glandulo'sus: glandular; full of glands

glaucifo'lius: with leaves that are gray; gray-green

glau'cus: covered with a bloom (that is, a fine white powdery coating such as graces the plum)

globo'sus: round; spherical

globulif'era: bearing globe-like structures

globulo'sus: like a little ball

glomera'tus: clustered; collected together in a head

glorio'sus: glorious; superb

gloxinioi'des: resembling *Gloxinia*

gluma'ceus: with glumes, that is, chaffy bracts that enclose the flowers of grasses

glutino'sus: glutinous; sticky

glycinioi'des: resembling the soybean, *Glycine max*

glycos'mus: sweet-smelling

gnaphaloi'des: resembling *Gnaphalium,* a genus of wooly-leaved herbs →

gomphoceph'alus: shaped like the head of a club or bolt

gongylo'des: knob-like; swollen

gonia'tus: angled; cornered

gonioc'alyx: with an angled calyx

gorgo'neus: Gorgon-like; snake-haired

gossyp'inus: resembling cotton, *Gossypium;* cottony

gracilen'tus: slender

graciliflo'rus: with slender or graceful flowers

gracil'ipes: with a slender stalk

grac'ilis: graceful; slender

gracil'limus: very slender

grae'cus: Greek; of or from Greece

gramin'eus: resembling grass; grass-like ↓

gramma'tus: striped by raised lines; inscribed

granaden'sis: of or from Granada, Spain; of or from
  Colombia, South America, formerly New Granada

gran'diceps: bearing large heads

grandicus'pis: with large cusps or points

grandidenta'tus: with large teeth

grandiflo'rus: with large flowers;
  free-flowering

grandifo'lius: with large leaves

grandifor'mis: on a large scale

grandipuncta'tus: with large spots

gran'dis: large; big; showy

granit'icus: growing on or in the
  crevices of granite

granula'tus: covered with minute grains

granulo'sus: composed of minute grains

RABBIT TOBACCO
Gnaphaloides. Gnaphalium
is a large genus of
wooly-leaved plants.
The Greeks used the
downy leaves for stuff-
ing small cushions and
pillows. *Gnaphalium
obtusifolium* is the rabbit
tobacco of the Ameri-
can South. For genera-
tions, naughty boys
have smoked the leaves.

❦

HIDDEN FLOWERS
Gramineus. "*Iris graminea*
I do possess, and grow
less for the garden value
of its pinky-mauve
flowers, almost hidden
at the base of the leaf-
clump, as for its curious
scent, like a sun-warmed
greengage. It is essen-
tially a little iris for
picking, and can be seen
very prettily marked
once you have sorted it
out from its leaves and
can observe it closely in
a glass or vase."
*V. Sackville-West's Garden
Book,* 1983

HERB OF GRACE ↑
Graveolens. "Ruta grave-olens. Rue. Herb of Grace. Very bitter tasting herb with grey leaves, strongly aromatic, native of the Mediterranean region. Once very important in domestic medical practice and believed to ward off witches and contagion. It was one of the important strewing herbs of old times. A few sprigs of Rue hung in the room will drive out flies." Louise Beebe Wilder, *The Fragrant Path,* 1990

gratiano'politanus: of Grenoble, France

grat'iola: agreeable (medicinally)

gratis'simus: very pleasing or agreeable

gra'tus: pleasing; agreeable

grave'olens: heavily scented; strong-smelling

gris'eus: gray

groenlan'dicus: of Greenland

grosse-serra'tus: with large sawteeth

gros'sus: very large; thick; coarse

grui'nus: resembling a crane

guianen'sis: of or from Guiana, South America

guineen'sis: of or from western Africa, the Guinea Coast

gummo'sus: gummy

gunneraefo'lius: resembling *Gunnera,* with big leaves

gutta'tus: spotted; speckled

gymnocar'pus: bearing naked fruits, that is, without any covering such as a perianth, hair, or mucus

gymnoceph'alus: bareheaded

gy'rans: revolving in a circle; gyrating

gyro'sus: spiral

abrotrich'us: covered with graceful or beautiful hairs

hadriat'icus: of or from the shores of the Adriatic

haeman'thus: with blood red flowers

haemato'calyx: having a blood red calyx

haemato'des: bloody; blood red in color

hakeoi'des: like *Hakea,* an Australian evergreen shrub

halimifo'lius: with soft leaves resembling *Atriplex halimus* (formerly *Halimium*)

haloph'ilus: growing in a salty habitat

hama'tus, hamo'sus: hooked at the tip; barbed

hamulo'sus: covered with small hooks; armed

harma'lus: responsive; sensitive

harpophyl'lus: with sickle-shaped leaves

hasta'tus: resembling a spearhead or arrowhead

hastif'era: with spear-shaped structures

hasti'lis: of a javelin or spear

hebecar'pus: producing down-covered fruit

hebephyl'lus: with downy leaves

hebra'icus, hebri'acus: Hebrew

hedera'ceus: of the ivy, *Hedera* →

hederifo'lius: with ivy-shaped leaves

hed'ys: sweet; having a pleasant taste or aroma

SUNFLOWERS
Helianthoides. "It has been erroneously supposed by poets, and others, that these flowers called sunflowers continually turn to the sun. Some flowers as *Anagallis . . .* and this *Helianthemum*, our English rock rose, expand the best in bright sunshine, while others have a different habit, as *Trogopon:*—'for it shutteth it selfe at twelue of the clocke' . . . Gerard."
Randal Alcock, *Botanical Names for English Readers*, 1876

🌠

ABODE OF SNOW
Himalaicus. Himalaya is from Sanskrit, *bima* for snow, *alaya* for dwelling. Many plants are commonly called Himalayan. Two are Himalayan pine, *Pinus gerardiana*, and Himalayan rhubarb, *Rheum nobile.*

helianthoi'des: resembling the sunflowers, *Helianthus*

helico'nia: of or from Mount Helicon, Greece; of the muses

hellenicus: of or from Greece

helo'des: of bogs

helvet'icus: Swiss

hel'vus: light bay to pale red in color; dingy

hemiphloe'us: half covered with bark

hemisphaer'icus: shaped like half a ball

hepat'ica: resembling a liver in color or shape; used in the treatment of liver disease →

hepaticaefo'lius: with leaves like liverwort, *Hepatica*

heptamer'us: in multiples of seven; with seven parts

heptaphyl'lus: with seven leaves (as in a group)

heracleifo'lius: with leaves resembling the cow parsnip, *Heracleum*

herba'ceus: not wood-forming; of the nature of an herb; low-growing; dying back to the ground annually

hermae'us: of or from Mount Hermes, Greece

hesper'ius: of the West; of the evening

heteracan'thus: with differing spines

heteran'thus: producing two or more kinds of flowers (as sterile and fertile) on one plant *heterocar'pus:* producing two or more kinds of fruit

heterodox'us: diverse, differing, or uneven

heteromor'phus: various in form; irregular in structure

heterophyl'lus: bearing leaves of more than one form

hexago'nus: six-angled

hexam'erus: in multiples of six; with six parts

hexan'drus: with six stamens

hi'ans: openmouthed; gaping

hiberna'lis: pertaining to winter

hiber'nicus: of or from Ireland (*Hibernia* in Latin)

hiber'nus: flowering or green in the winter; of Ireland

hibiscifo'lius: with leaves like *Hibiscus* ↘

hiema'lis: of the winter; winter-flowering

hierochun'ticus: of Jericho, from the classical name of Jehrico

himala'icus: Himalayan

hippoman'icus: with the ability to drive horses mad

hirci'nus: smelling like a goat

hirsu'tulus: somewhat hairy

hirsu'tus: hairy; covered with coarse hairs

hir'tipes: with hairy stalks or stems

hir'tus: hairy

hispalen'sis: of or from Seville or southern Spain

COARSE THINGS
Herbaceus. "Plaintive letters reach me from time to time saying that if I do not like herbaceous borders what would I put in their place? It is quite true that I have no great love for herbaceous borders or for the plants that usually fill them—coarse things with no delicacy or quality about them. I think the only justification for such borders is that they shall be perfectly planned, both in regard to colour and to grouping; perfectly staked; and perfectly weeded. How many people have the time or the labour? The alternative is a border largely composed of flowering shrubs, including the big bush roses."
V. Sackville-West's Garden Book, 1983

SNOWFLAKE SEASON ↑
Hyemalis. "The generic
name of the snowflake,
Leucojum, is from the
Greek. It means white
violet, and was given to
the plant because of
the fragrance of the
flowers. I never knew
that they were fragrant
until I read it in a
book. The perfume is
so subtle that you
must warm the flowers
in your hand before
you become aware of
it. If you trust in

hispan'icus: Spanish

his'pidus: bristly; coarse; with stiff hairs

hollan'dicus: of or from Holland; of or from northern New Guinea

holocar'pus: bearing undivided fruits

holochry'sus: wholly golden

holoseric'eus: completely covered with silky hairs; having a silky sheen

holos'tea: whole bone; the Greek name for an herb resembling chickweed

homocar'pus: with only one kind of fruit

homo'lepis: covered with uniform scales

horizonta'lis: horizontal; flat on the ground; spreading horizontally

hor'ridus: prickly; very thorny; bristly

horten'sis, horto'rum, hortula'nus, hortula'lis, hortulo'rum: belonging to a hortus or garden, or to gardens; cultivated

humifu'sus: sprawling on the ground; procumbent

humilifo'lius: with leaves like the hop, *Humulus*

hu'milis: low-growing; dwarf

hyacinth'inus: deep purplish blue in color

hyacinthoi'des: resembling a hyacinth, *Hyacinthus*

hyal'inus: transparent; translucent; colorless

hyberna'lis: pertaining to the winter

hyb'ridus: mixed; mongrel; crossbred

hydrangeoi'des: resembling *Hydrangea*

hydroph'ilus: water-loving; growing on or in water; needing water for pollination

hyema'lis: of winter

hygromet'ricus: taking up water

hylae'us: belonging to the woods

hymenan'thus: flowers with or covered by a membrane

hymeno'des: resembling a membrane

hymenorrhi'zus: with membranous roots

hyperbo'reus: of or from the far North

hypericifo'lius: with leaves like St. Johnswort, *Hypericum*

hypericoi'des: resembling St. Johnswort, *Hypericum*

hypnoi'des: resembling the mosses, once grouped as *Hypnum*

hypochondri'acus: of melancholy or somber appearance

hypogae'us: underground; growing or developing underground

hypoglau'cus: glaucous beneath

hypoleu'cus: whitish or pale beneath, as a leaf

hypophyl'lus: under or beneath the leaf

hypopith'ys: growing under pines

hyrca'nium: from Hyrcania, by the Caspian Sea

hyssopifo'lius: with leaves like hyssop, *Hyssopus* →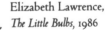

hys'trix: porcupine-like, bristly, or spiny

names there is a snowflake for each season—winter, summer, autumn, spring—but, like so many ideas that are charming in the abstract, the naming does not work out so well. The winter snowflake, *Leucojum hyemale,* blooms in spring; the summer snowflake, *L. aestivum,* often blooms in winter in these parts; and the autumn snowflake, *L. autumnale,* blooms in summer."
Elizabeth Lawrence, *The Little Bulbs,* 1986

Idaeus. "Meanwhile Hera rapidly drew near to Gargarus, the highest crest of lofty Ida. Zeus the Cloud-compeller saw her, and at the first look his heart was captured by desire, as in the days when they had first enjoyed each other's love and gone to bed together without their parents' knowledge." Homer, *Iliad*

Mount Ida was abuzz with immortal love. Homer has Hera seducing Zeus there so that Poseidon can lead the Greeks into battle against the Trojans. Hera lost out in love another time on Mount Ida when Paris, who had been abandoned on the slopes as an infant and grown up there, spurned Hera—and Athena—for Aphrodite's offer.

ad'inus: jade green in color

ian'thinus: violet, violet-blue in color

iber'icus, iberid'eus: of the Iberian Peninsula, Spain and Portugal; or from the Caucasus, Georgia

iberidifo'lius: with leaves like candytuft, *Iberis*

icosan'drus: with twenty stamens

icter'icus: yellowed; jaundiced

idae'us: of Mount Ida, Turkey; or of Mount Ida, Crete

ignes'cens: fiery red in color

ig'neus: fiery red in color

ikar'iae: of or from the Aegean island Nikaria

ilicifo'lius: with holly-like leaves →

illecebro'sus: of the shade; alluring; enticing

illini'tus: smeared upon; smudged

illinoinen'sis: of or from Illinois, United States

illustra'tus: as if painted upon

illus'tris: bright; brilliant

illyr'icus: of Illyria, near the Dalmatian coast

ilven'sis: of or from the island of Elba; of or from the river Elbe, Bohemia and Germany

imbecil'lis: feeble; weak

imber'bis: without beard, spines, or hairs

imbrica'tus: overlapping, as shingles on a roof

immacula'tus: unblemished; spotless

immargina'tus: without a distinct margin or border

immer'sus: growing underwater; submerged

impa'tiens: impatient; throwing seed when ripe

impera'tor: commanding; outstanding

imperatri'cis: of the Empress Josephine ↘

imperia'lis: imperial; royal; showy

implex'us: interwoven; tangled; interlaced

impres'sus: sunken in; furrowed; below the surface

impud'icus: shameful; lewd; impudent

inaequa'lis: unequal

inaequilat'erus: having unequal sides

inaper'tus: closed; not open

inca'nus: hoary; light gray

incarna'tus: flesh-colored; pink

incer'tus: uncertain; doubtful

inci'sus: deeply and irregularly cut

inclau'dens: never closing

inclina'tus: bent downward; below the horizontal

incomparab'ilis: incomparable; excellent

incomp'tus: rude; unadorned

inconspic'uus: small

incrassa'tus: thickened toward the summit

HOLIDAY HOLLIES
Ilicifolius. The spiny, glossy evergreen leaves with their red berries of the English holly (*Ilex aquifolium*) are what we think of at Christmastime, but there are many hollies with other foliage and fruits. The leaves of *I. decidua* are gone by Christmas, but only to better reveal a grand show of red to orange berries. Virgil's holly in the *Georgics,* though, is not a holly at all. It is the holly oak, *Quercus ilex.*

❧

NAPOLEONIC ROSES
Imperatricis. Most of the plants carrying the epithet *imperatricis* refer to one empress, Josephine (1763–1814). Her fabled collection of roses at Malmaison—the largest ever assembled—is now being reconstructed.

❧

EAST IS EAST
Indicus. Rosa indica originated near Canton, China, and indicates the loose usage of the term *indica.* Plants native throughout the Far East Indies and as far away as Japan were labeled as *indicus. Canna indica* is native to tropical America, *Rhododendron indicum* from Japan, and so on.

❧

WINTER ORCHIDS
Insignis. Paphiopedilum insigne is one of the most easily grown orchids for winter bloom. Vita Sackville-West agrees. She says a greenhouse isn't necessary; "a window-sill in a room where the temperature never dropped below 45 degrees should suffice. They do not like very strong sunlight, and they do like plenty of water."

❧

incur'vus: bent inward; inflexed

indehis'cens: not opening at all; not splitting when ripe

indenta'tus: indented

in'dicus: of or from India or the Far East

ine'brians: intoxicating

iner'mis: unarmed; without thorns

infecto'rius: used for dyes

infes'tus: dangerous; troublesome; becoming a weed

infla'tus: swollen; filled

infortuna'tus: unfortunate; often poisonous

infrac'tus: curved inward; sharply bent

infundibulifor'mis: funnel-form; trumpet-shaped

in'gens: enormous; exceeding the expected size

inodo'rus: scentless

in'quinans: blemished; dirty; polluted

inscrip'tus: as if marked with letters

insectif'era: bearing insects; with insect-shaped flowers

insig'nis: remarkable; distinguished; marked ↗

insitit'ius: grafted

insula'ris: pertaining to or growing on islands

intac'tus: untouched; unopened

in'teger: entire; undivided

integrifo'lius: with entire or uncut leaves

interjec'tus: cast between; intermediate in form

intertex'tus: interwoven; intertwined

intor'tus: twisted; bent upon itself

intrica'tus: entangled

intror'sus: turned inward; turned toward the axis

intumes'cens: swollen; puffed up

intyba'ceus: resembling chicory

inunc'tus: anointed; glossy as if oiled

inunda'tus: flooded; growing in the flood plains

invi'sus: unseen; overlooked; not visible

involucra'tus: with an involucre, that is, a ring of bracts
    surrounding the flowers

involu'tus: rolled inward; obscured

io'des: resembling *Viola* ↘

ionan'thus: with violet-colored flowers

ir'icus: Irish

iridiflo'rus: with flowers like *Iris*

irrig'uus: well-watered; wet

irrit'ans: irritating; causing a rash

isabelli'nus: dirty yellow; tawny

isan'drus: with equal stamens

isopet'alus: with equal or similar petals

is'tria: of Istria, Yugoslavia

ital'icus: Italian

iteophyl'lus: willow-leaved

ixocar'pus: sticky-fruited

THE BLUE FLAG

Iridiflorus. "All my heart became a tear, ∾ All my soul became a tower, ∾ Never loved I anything ∾ As I loved that tall blue flower! ∾ It was all the little boats ∾ That had ever sailed the sea, ∾ It was all the little books ∾ That had gone to school with me; ∾ On its roots like iron claws ∾ Rearing up so blue and tall— ∾ It was all the gallant Earth ∾ With its back against the wall!" Edna St. Vincent Millay, "The Blue Flag in the Bog," 1921

❀

HIDDEN MEANINGS

Isabellinus. The tawny color that *isabellinus* implies comes from the spurious story that the Archduchess Isabella did not change or wash her underwear for three years.

❀

MEXICAN HEAT

Jalapa. Jalapa, Mexico, is the home of the fiery jalapeño pepper, which is popularly identified with Mexican food. Diana Kennedy in *The Cuisines of Mexico* says they are also called *chiles gordos* —fat chilies—and *cuaresmeno.* Dried and smoked they are called *chiles chipotles.* In the Park seed catalog of 1989, we find American hybridists turning down the heat. With "Tam Mild Jalapeño . . . Now enjoy all the extraordinary flavor of Jalapeño Pepper without all the heat. Only mildly pungent."

acobae'us: honoring Saint James (Jacobus); of or from Saint Iago, Cape Verde Islands

jala'pa: of or from Jalapa, Mexico

jamaicen'sis: of or from Jamaica

japon'icus: Japanese

jasmin'eus: like jasmine, *Jasminum*

jasminiflo'rus: jasmine-flowered

javan'icus: of Java

jeju'nus: meager; small

johan'nis: of or from Port Saint John, South Africa

jonquil'la: with rush-like leaves

juanen'sis: from Genoa, Italy

juba'tus: crested; with a mane

jucun'dus: agreeable; pleasing

jugo'sus: joined; yoked together

julifor'mis: downy

jun'ceus: resembling a rush

juncifo'lius: with leaves like the rush

juniperifo'lius: with juniper-like leaves

juniperi'nus: resembling *Juniperus;* bluish brown, like berries of junipers ↗

a'ki: from *kaki-no-ki,* Japanese for the persimmon ⟍

kalmiaeflo'rus: with flowers like the mountain laurel, *Kalmia*

kamtschat'icus: of or from the Kamchatka peninsula, Russia

kansuen'sis: of or from Kansu in northwest China

karatavien'sis: of or from the Kara Tau mountains in Kazakhstan, central Asia

kashmiria'nus: of Kashmir, south Asia

kele'ticus: charming

kermesi'nus: carmine-colored; purplish red

kewen'sis: in honor of the Royal Botanic Gardens at Kew, England

kiusia'nus: of or from Kyushu, a southern island of Japan

kolomic'ta: colloquial name for an *Actinidia* from the Amur River, Manchurian border, China

korea'nus, koria'nus, koraien'sis: of Korea

kou'sa: Japanese for a species of dogwood, *Cornus*

kurdi'cus: of or from the land of the Kurds, western Asia

KALM'S LAUREL

Kalmiaeflorus. Kalmia is the genus of what Southerners call mountain laurel, ivy, mountain ivy, sheep laurel, and wicky. Botanically, it is named for Peter Kalm, a student of Linnaeus who collected in the Southern states in the eighteenth century.

❧

ROCK GARDEN GEMS

Kansuensis. Allium kansuense. "This is one of the charming Blue Garlics about which confusion seems to cling . . . a gem for the rock garden by whatever name we have it, and especially important because of the sparse-flowered season at which it blooms among our small hills."

Louise Beebe Wilder, *Adventures with Hardy Bulbs,* 1989

❧

AN ENGLISH WALK
Laburnifolius. An arbor
or arch of *Laburnum*
was a typically English
walk from the begin-
ning of the century.
Gertrude Jekyll
described one in *Colour
Schemes for the Flower
Garden:* "It passes down
through the middle of
the kitchen garden and
is approached by an
arch of laburnum. It
is backed on each side
by a Hornbeam hedge
some five and a half
feet high." And anoth-
er English writer,
Christopher Lloyd,
said another century
later in *The Well-
Tempered Garden,* "My
friend said, 'I have just
planted another labur-
num. They tell me it is
suburban, but I adore
laburnums.' I was
whole-heartedly on
his side."

L*abia'tus:* with a well-developed lip

lab'ilis: slippery; unstable; perishable

laburnifo'lius: resembling the golden chain tree, *Laburnum* ↓

lac'erus: roughly torn; with a rough fringe

lacinia'tus: torn; jagged; slashed into long narrow pieces

lac'rimans: weeping; causing tears

lac'ryma-jobi: Job's-tears

lacta'tus: milky

lac'teus: milk white

lactif'era: with a milky sap

lactiflo'rus: bearing milk-colored flowers

lacuno'sus: with holes or pits

lacus'tris: pertaining to lakes

ladanif'era, ladan'ifer: yielding the opiate laudanum

laetes'virens: bright or vivid green

laetiflo'rus: bright- or pleasing-flowered

lae'tus: bright; vivid

laevicau'lis: smooth-stemmed

laeviga'tus: smooth; as if polished

lae'vis: smooth; free from hairs or roughness

lagena'rius: shaped like a bottle or flask

lago'pus: hare-footed ⬎

lamella'tus: arranged in layers

lana'tus: wooly

lanceola'tus: armed with a lance or spear point

lan'ceus: spear-shaped

lanig'era: wool-bearing; wooly

lano'sus: wooly

lanugino'sus: wooly or downy

lappa'ceus: with small burs

lappon'icus: of Lapland

laric'inus: resembling the larch, *Larix*

lasian'drus: with downy stamens

lasian'thus: wooly-flowered

lasiocar'pus: with rough or wooly fruits or cones

lasiopet'alus: with downy petals

latebro'sus: growing in dark or shady places

lateriflo'rus: flowering on the side

laterit'ius: dark brick red in color

latifo'lius: broad-leaved

lat'ifrons: broad-fronded

latila'brus: with a broad or wide lip

latimacula'tus: with broad spots

la'tus: broad; wide

lauda'tus: lauded; praiseworthy

MAGICAL TEARS
Lacryma-jobi. From Elizabeth Lawrence's *Through the Garden Gate:* "Corn beads or 'Indian beads for stringing on string,' as they were described in an advertisement by Mrs. W. L. Null, who wrote me in response to my inquiry: 'Concerning the bead seeds, yes, I grow them. When they come up they look just like corn, and will sucker out from the mother stalk. They do not make a bloom. Later on you will notice beads in the buds of the limbs all over the bush. When they turn brown it is time to gather them and spread them out before stringing.' Indian Beads are listed in the Park catalog as Job's-tears, *Coix lacryma-jobi.* . . . They are said to be magic."

❦

LAVENDER SWINDLES *Lavandulaceus.* "Around my rose beds are some sixty lavenders, all ordered and planted at the same time, all sold to me as 'true English lavender.' Even as I dug them in I saw the little plants differed from one another, but I put this down to their immaturity. I was wrong. The passage of time confirmed that I had several varieties and was once more adrift in the sea of herbal confusion. Some had blue-green foliage, others grey; one plant had leaves like needles; its neighbor's were broader and flatter; the colors varied from white to deep purple and the heights from one to three feet. Had I been swindled? Why were they so different? Did it matter?" Eleanor Perenyi, *Green Thoughts,* 1981

laurica'tus: wreathed; resembling laurel

lauri'nus: laurel-like

lau'tus: washed

lavandula'ceus: resembling lavender, *Lavandula*

lax'us: open; loose; flaccid; loosely arranged

laz'icus: from Lazistan, northeastern Turkey

lebetifor'mis: basin-shaped

ledifo'lius: with leaves like Labrador tea, *Ledum*

leiocar'pus: producing smooth fruits

leiog'ynus: having a smooth pistil

leiophyl'lus: with smooth leaves

lenticula'ris: lens-shaped

lentigino'sus: freckled

len'tus: pliant; tenacious; tough

leo'nis: colored or toothed like a lion

leonu'rus: like a lion's tail

leopardi'nus: spotted like a leopard

lepidophyl'lus: scaly-leaved

lepido'tus: covered with small scales

lep'idus: graceful; elegant

lepro'sus: scurfy; spotted like a leper

leptocau'lis: with a thin, graceful stem

leptol'epis: covered with thin scales

leptophyl'lus: with slender or graceful leaves

leucanthemifo'lius: with foliage like the oxeye daisy ↑

leucan'thus: with white flowers

leuco'botrys: with white clusters

leucocau'lis: white-stemmed

leucophae'us: dusky white in color

leucoxan'thus: whitish yellow

leuco'xylon: having white wood

libanot'icus: of Libania; of Mount Lebanon

libe'ricus: of or from Liberia, western Africa

libur'nicus: of Liburnia, the Adriatic coast of Croatia

liby'cus: of or from Libya, northern Africa

lignatil'is: growing on wood

ligno'sus: woody

ligula'ris, ligula'tus: like a strap

ligus'ticus: of Liguria, formerly Ligusticum

ligustrifo'lius: with leaves like privet, *Ligustrum*

ligustri'nus: resembling privet, *Ligustrum*

lilac'inus: lilac in color

lilia'ceus: resembling the lily family, *Liliaceae* ↗

liliiflo'rus: with lily-like flowers

lilliputia'nus: of very small or low growth; Lilliputian

limba'tus: bordered; marked by a margin

limen'sis: of or from Lima, Peru

limno'philus: swamp-loving

limo'sus: of muddy or marshy places

RAIN IN LIBYA
Libycus. "I have now mentioned all the Libyans whose names I am acquainted with; most of them cared nothing for the king of Persia, either then or now. . . . I do not think of the country for the fertility of its soil with either Asia or Europe, with the single exception of the region called Cinyrps—so named after the river which waters it. This region, however, is quite different from the rest of Libya, and is as good for cereal crops as any land in the world. The soil here, unlike the soil elsewhere, is black and irrigated by springs; it has no fear of drought on the one hand, or of damage, on the other, from excessive rain (it does, by the way, rain in that part of Libya)." Herodotus, *Histories*

MODEST LINNAEUS
Linnaeanus. The genus
Linnaea was named by
and for himself, Carl
Linnaeus (1707–1778).
His binomial system
for naming plants
revolutionized all
intercontinental horti-
cultural communica-
tion, but his comment
on the twin-flower is
typically humble: "a
plant of Lapland,
lowly, insignificant,
disregarded, flowering
but for a brief space—
from Linnaeus who
resembles it."

❦

YELLOW DUST
Lotoides. "Through
every hollow cave and
alley lone ∾ Round
and round the spicy
downs the yellow
Lotos-dust is blown."
Alfred, Lord Tennyson,
"The Lotus-Eaters,"
1833

❦

limpi'dus: clear; transparent; without color

linarioi'des: resembling toadflax, *Linaria*

linearifo'lius: linear-leaved

linea'ris: narrow; with nearly parallel sides

linea'tus: lined; with lines or parallel stripes

linguefor'mis: tongue-shaped

linifo'lius: with leaves like flax, *Linum*

linnaea'nus: in honor of Linnaeus

linoi'des: resembling flax, *Linum* ↘

litangen'sis: of or from Litang, Szechuan, China

litho'philus: rock-loving; growing on rocks

lithosper'mus: producing very hard seeds

litien'sis: of or from Litiping, Yunnan, China

littora'lis: of the seashore

lituiflo'rus: trumpet-flowered

liv'idus: lead-colored; bluish gray in color

loba'tus: with lobes

lobelioi'des: resembling *Lobelia*

lobula'ris: lobed

lobula'tus: with small lobes

loch'imus: growing in thickets

locus'tus: with small spikes; thorny

lolia'ceus: like rye grass, *Lolium*

longicau'lis: long-stemmed

longifo'lius: long-leaved

longi'labris: long-lipped

long'ipes: long-stalked

lon'gus: long

lophan'thus: with crested flowers

lora'tus: strap-shaped

lorifo'lius: with strap-like leaves

lotoi'des: like a lotus ↘

louisia'nus: of or from Louisiana, United States

lu'cidus: glittering; shining; clear

luco'ris: of the woods

luna'tus: crescent-shaped

lunula'tus: somewhat crescent-shaped

lupuli'nus: resembling hops, *Humulus*

lu'ridus: sallow; wan; pale yellow

lusitan'icus: of or from Portugal, formerly Lusitania

luta'rius: growing in mud or muddy places

lutetia'nus: of or from Paris, formerly Lutetia

lu'teus: yellow

luxu'rians: luxuriant

lychnidifo'lius: with wooly leaves like *Lychnis*

lyci'us: of or from Lycia, southern Turkey

lycoc'tonum: wolf-poison

lycopodioi'des: resembling club moss, *Lycopodium*

lyra'tus: lyre-shaped; broadly rounded toward the tip

lysolep'sis: with loose scales

MOUNTAIN TREE
Luteus. Cledrastis lutea.
"The showy racemes
of this native tree are
8–24 inches or more
long. The gray or light
brown bark of the
trunk is smooth: the
wood is yellow and has
been used as a source
of dye. The tree of the
Appalachian area is
found in the rich
woods of only a few
of our mountain
counties. It is often
planted elsewhere as
an ornamental."
Justice and Bell,
*Wild Flowers of North
Carolina,* 1987

ARISTOTLE'S PUPIL
Macedonicus. Macedonia

ARISTOTLE'S PUPIL
Macedonicus. Macedonia
was a kingdom north
of Greece of small
notice until the con-
quests of Phillip, the
father of the Great
Alexander. Alexander's
tutor was Aristotle,
and in 326 B.C. Alexan-
der ascended the
throne at the death of
his father. By 330 B.C.,
Alexander had essen-
tially conquered the
ancient Persian Empire
and set up the hellenis-
tic structure of art,
philosophy, and poli-
tics that touched all
ancient Mediterranean
culture.

❦

FLOWERS OF MAY
Majalis. "O gallant
flowering May, ∾
Which month is
painter of the world,
∾ As some great
clerks do say."
Every people has
selected its own flower
to celebrate May Day,

acedon'icus: Macedonian

macel'lus: lean; meager

macracan'thus: with large thorns or spines

macran'thus: producing large flowers

macula'tus: spotted; blotched

madagascarien'sis: of or from Madagascar

maderaspat'anus: of or from the region of Madras, India

maderen'sis: of or from Madeira, a group of islands off the
  coast of Morocco

maesi'acus: of Moesia (Bulgaria and Serbia)

magellan'icus: of or from the Straits of Magellan

magellen'sis: of or from Mount Majella, Italy

mag'nus: large

maja'lis: of May; flowering in May

ma'jor, ma'jus: greater; larger

malabar'icus: of or from the Malabar coast,
  southern India

malacoi'des: soft; mucilaginous; tender; weak

malifor'mis: shaped like an apple →

malva'ceus: resembling the mallow, *Malva*

malvi'nus: mauve in color; resembling the mallows

mammilla'ris, mammo'sus: with breast- or nipple-like
  structures

mandshu'ricus, mandschu'ricus: of Manchuria, China

manica'tus: sleeved; having an easily removed surface

manzani'ta: little apple

margarita'ceus: pearl-like; with the sheen of pearls

margina'lis: with a distinct margin or border

maria'nus, marilan'dicus, marylan'dicus: of or from Maryland, United States

marin'us, marit'imus: growing by or in the sea

marmora'tus, marmo'reus: marbled; mottled

marocca'nus: of or from Morocco, northern Africa

marta'gon: resembling a Turkish turban ↘

martinicen'sis: of or from Martinique, West Indies

mas', mascula'tus, mas'culus: male; masculine; bold

matricariaefo'lius: with leaves like false chamomile, *Matricaria*

matrona'lis: pertaining to the Roman festival of the matrons held on the first day of March

mauritan'icus: of Mauritania, northern Africa

mauritia'nus: of the Mauritius Islands, Indian Ocean

maxilla'ris: shaped like the upper jaw

max'imus: largest

mays': native American name for corn

med'icus: useful as a medicine; curative; of Media (Iran)

mediopic'tus: striped down the center

mediterra'neus: of the Mediterranean region

me'dius: in-between in size or shape

the day that old writers saw as the death of winter and the birth of summer. Sage belongs to May. Old sayings point to its longevity: "Set sage in May, and it will grow alway" and "Why should a man die while sage grows in his garden?" But more than a hundred flowers are called Mayflower. In England, the name is passed among hawthorns, marsh marigolds, lilacs, snowballs, and cowslips.

HEN'S PLUMAGE
Meleagris. Fritillaria melea-gris. "Snake's Head Fritillary, Guinea Hen Flower, Checkered Lily. . . . The Checkered Lily was called by Dodonaeus—the sixteenth century botanist of Flanders, whose writings are said to form the basis of Gerard's *Herbal*—*Flos meleagris, meleagris* then being the name of the guinea hen, for the reason that the whole flower is checkered over like the wings and breast of that curious fowl. 'Nature, or rather the Creator of all things, hath kept a very wonderful order, surpassing (as in all other things) the curiousest painting that art can set downe. One square is of a greenish yellow colour, the other purple, keeping the same order on the backside of the flower as on the

medulla'ris: soft-wooded; pithy

megacar'pus: producing large fruits

megalan'thus: producing large flowers

megapotam'icus: of the big river

megarrhi'zus: with large roots

melanchol'icus: sad-looking, hanging, or drooping

melanocau'lon: with black stems

melanocen'trus: black at the center

melanococ'cus: black-berried

melanoleu'cus: black and white

mel'anoxylon: with black wood

melea'gris: like a guinea-fowl, speckled or checkered ↑

meliten'sis: of or from the Mediterranean island of Malta

mel'leus: pertaining to honey

mellif'era: honey-bearing

melli'tus: sweet as honey

melofor'mis: shaped like a melon

membrana'ceus: skin-like; very thin

meniscifo'lius: having crescent-shaped leaves

meridiona'lis: southern; pertaining to or flowering at noon

mesoleu'cus: with a white central stripe

mexica'nus: Mexican

micace'us: growing in mica-laden soils

mi'cans: glittering; sparkling; shiny

microceph'alus: forming small heads

microm'eris: with few or a small number of parts

micropet'alus: with small petals

microste'mus: composed of small filaments

microthe'le: resembling a small nipple

mikanioi'des: resembling the climbing hempweed, *Mikania*

milia'ceus: pertaining to millet ↘

milita'ris: soldier-like; of or belonging to a soldier

millefolia'tus, millefo'lius: with a thousand leaves; a leaf cut into a thousand parts

mime'tes: mimicking

mimosoi'des: resembling *Mimosa*

mi'mus: mimic; farce

mi'nax: menacing; forbidding

minia'tus: cinnabar red; vermilion

min'imus: least; smallest

mi'nor, mi'nus: smaller

minu'tus: minute; very small

mirab'ilis: marvelous; astonishing

missourien'sis: of or from Missouri, (the state or the river) United States

mi'tis: mild; gentle; ripe; without thorns

mitra'tus: shaped like a turban or mitre

mix'tus: mixed

moesi'acus: from the lower Danube, formerly called Moesia, now part of Bulgaria

inside, although they are blackish on one square, and of a violet colour in another: in so much that every leaf seemeth to be the feather of a Ginnie hen, whereof it took its name.'"
Louise Beebe Wilder, *Adventures with Hardy Bulbs*, 1989

❧

BEEMAN'S GARDENS *Melleus.* "Let there be ❧ Gardens to tempt them, breathing saffron flowers, ❧ And, stationed with willow sickle to scare off thieves ❧ and birds, Priapus of the Hellespont. The beekeeper must go himself to fetch ❧ Thyme from the mountain heights and laurestines ❧ For growing round the hive, himself must harden ❧ His hands with rugged work."
Virgil, *Georgics*

❧

ORNAMENTS OF MAY ↑
Morifolius. "The Mulberry Tree [*Morus*] every-where amidst the Woods grows wild: The Planters, near their Plantations, in Rows and Walks, plant them for Use, Ornament, and Pleasure: What I observed of this Fruit was admirable; the Fruit there, was full and ripe in the latter end of April and beginning of May, whereas in England and Europe, they are not ripe before the latter end of August."
Thomas Ashe,
Carolina, 1682

moldav'icus: of or from Moldavia, eastern Europe

mol'lis: soft; flexible; mild

molucca'nus: of or from the Moluccas, Indonesia (also known as the Spice Islands)

monacan'thus: with one spine or thorn

monadel'phus: having stamens in bundles

monen'sis: of or from the Isle of Man, formerly Mona

mongol'icus: of or from Mongolia

monilifor'mis: necklace-like; strung like beads

monoi'cus: with male and female flowers on the same plant

monopyre'nus: a fruit forming one hard stone or nut

monspessula'nus: of or from Montpellier, France

monstro'sus: monstrous; abnormal in size or shape

monta'nus: growing in the mountains

montic'olus: inhabiting mountains

montig'enus: originating in the mountains

morifo'lius: having leaves like the mulberry, *Morus*

mosa'icus: colored like a mosaic

moscha'tus: musky scent

muco'sus: slimy

mucrona'tus: with a sharp point or edge

multi'cavus: with many hollows; pitted

mul'ticeps: forming many heads

multicosta'tus: with many ribs; many-sided

multif'idus: divided or parted frequently

multifurca'tus: forked or divided often

multi'jugus: many parts joined

mul'tiplex: folded frequently

mu'me: Japanese for the flowering apricot

mun'dus: trim; neat; elegant; handsome

muni'tus: armed; fortified

mura'lis: growing on walls

murica'tus: roughened on the surface by means of
   sharp, hard points

musa'icus: resembling the banana plant, *Musa*

muscaetox'icum: fly-poison

muscip'ula: fly-catcher →

musciv'orus: fly-eating

muscoi'des: resembling the mosses

musco'sus: mossy; with a moss-like surface

mutab'ilis, muta'tus: fickle; changeable

mu'ticus: blunt; without a point

myriacan'thus: with many spines

myrio'cladus: with many branches

myrmecoph'ilus: ant-loving

myrtifo'lius: with leaves like myrtle, *Myrtus*

mystaci'nus: moustached

myu'ros: with a mouse tail

FLY POISON
Muscaetoxicum. "Fly-poison. *Amianthium muscaetoxicum.* The dense raceme is usually 2–4 inches long on a slender stalk 12–24 inches tall. The white petals and sepals do not wither after the flower has been pollinated but persist on the plant and turn green as they age. As the name implies, the plant is poisonous, especially the bulb." Justice and Bell, *Wild Flowers of North Carolina,* 1987

HOLY MOSSES
Muscoides. Thoreau wrote: "The beauty there is in mosses will have to be considered from the holiest, quietest nook."

SELF-ADORATION
Narcissiflorus. Narcissus
was the beautiful son
of Cephissus and the
nymph Leiriope, inces-
santly adored, but
never able to return
emotion. The nymph
Echo was among the
first to be spurned by
Narcissus; she pined
for him until only her
voice remained. When
Artemis finally heard
the cry of Ameinius
who had killed himself
on Narcissus's thresh-
old, she revealed Nar-
cissus's reflection in a
limpid pool, and
allowed him to fall in
love with himself.
When she denied him
consummation, he
stabbed himself by
the pool of water, and
the flower *Narcissus*
sprang from the
blood-soaked ground.
Robert Graves thinks
the name *Narcissus*
originally applied to
a blue fleur-de-lis.

anel'lus: very dwarf

na'nus: dwarf

napaulen'sis: of or from Nepal

napifor'mis: shaped like a turnip

narbonen'sis: of or from Narbonne, France

narcissiflo'rus: with flowers like *Narcissus* ↘

narino'sus: broad-nosed

nasu'tus: large-nosed

na'tans: floating; swimming

nauseo'sus: nauseating

navicula'ris: boat-shaped

neapolita'nus: of or from Naples, Italy

nebulo'sus: clouded-over; obscure; cloud-like

neglec'tus: overlooked; disregarded; slighted

nelumbifo'lius: with leaves like the lotus,
     Nelumbio

nemora'lis, nemoro'sus: of the woods; sylvan;
     covered with foliage

nepalen'sis: from Nepal; Nepalese

nepetoi'des: resembling catnip, *Nepeta*

nephrol'epis: with kidney-shaped scales

nereifo'lius, neriifo'lius: with leaves like oleander, *Nerium*

nervo'sus: prominently veined; sinewy

nicaren'sis: of or from Nice, France

nic'titans: blinking or moving; as if winking

ni'dus: nest-like

ni'ger: black in color; ominous; unlucky

nig'ricans: black

nigricor'nis: black-horned

nig'ripes: black-footed

nikoen'sis: of or from Nike, Japan

nilia'cus: of or from the Nile Valley

nilot'icus: of or from the Nile Valley; Egyptian

nippon'icus: of or from Nippon (Japan)

ni'tens, nit'idus: shining; glossy; as if polished

niva'lis, niv'eus: snowy; white; growing in snow

nob'ilis: noble; well-known; outstanding ↑

noctiflo'rus: night-flowering

noctur'nus: of the night

nodiflo'rus: flowering at the nodes or joints

nodo'sus: with nodes; jointed; conspicuously knotty

nodulo'sus: with small nodes

noli-tan'gere: do not touch; with fruit that bursts upon touch

nonscrip'tus: unmarked; without lines

norveg'icus: of or from Norway

BLACK ROOTS

Niger. Helleborus niger.
The *Hellebores* are one of the first discoveries gardeners make as they extend their gardening year-round. There are forms supplying flowers from Christmas to Easter, even in icy weather. The Christmas rose is *H. niger,* with white blooms from a black root. It is a highly poisonous, but effective, drug used since ancient days, recommended by Robert Burton for dispelling the effects of melancholia: "Borage and Hellebor fill two scenes, Sovereign plants to purge the veins of melancholy . . ." These are long-lived plants, resenting movement. In the English cottage tradition, the Christmas rose belongs at the front door.

🌿

TRUMPETS AT NIGHT
Nyctagineus. "One hot summer evening when I had gone to bed early and could not get to sleep, I slowly became aware of the most intoxicating odors, of lemon and honey and something else that is strong but evasive. The next morning I parted the ivy on the fence, and saw that it was the great white trumpets of Mrs. Dooley's datura that had been pouring elixir into the night. In the morning they are scentless." Elizabeth Lawrence, *Through the Garden Gate,* 1990

❧

nota'tus: marked or stamped; spotty; distinguished

no've-an'gliae: of New England, United States

no'vae-caesar'eae: of New Jersey, United States

no'vae-zealand'ia: of New Zealand

no'vi-bel'gii: of New York, United States

nubic'olus: dwelling among clouds

nu'bicus: of or from the Sudan, northern Africa (Nubia)

nucif'era: producing nuts

nuda'tus: naked; stripped; bare

nudiflo'rus: nude at flowering; flowering before leaves emerge

nu'dus: nude; naked; empty; bare

numid'icus: of Numidia, now Algeria, northern Africa

numis'matus: coin-shaped

nummula'rius: resembling coins or money

nu'tans: nodding; swaying to and fro

nyctagin'eus, nyctic'alus: night-blooming

nymphoi'des: resembling the water-lily, *Nymphea* →

nysae'us: of or from Mount Nysa, Thrace, the mythical birthplace of Bacchus (Dionysius)

bcon'icus: like an inverted cone

obe'sus: fat; swollen

obfusca'tus: clouded over; confused; indistinct

obla'tus: oval in shape; rounded at the ends

obli'quus: slanting; with unequal sides

oblitera'tus: obliterated; erased; wiped out; canceled

oblonga'tus, oblon'gus: oblong; elongated ↓

obova'tus: egg-shaped, with the narrower end down;
   inverted ovate

obscu'rus: hidden; unknown; dark and gloomy

obsole'tus: out-of-date; neglected; shabby

obtec'tus: covered up; protected

obtusa'tus: dull; blunt; rounded

obvalla'tus: walled up; enclosed

occidenta'lis: western; pertaining to
   the setting sun

occul'tus: hidden; secret

ocean'icus: growing near the ocean

ocella'tus: with small eyes; eye-like

ochra'ceus: reddish yellow

ochrea'tus: with a legging; having
   a tubular sheath

ochroleu'cus: yellowish white; buff

FORBIDDEN FRUIT
Oblongus. Cydonia oblonga,
common quince. The
quince was once
thought to be the for-
bidden fruit of the
Garden of Eden. In
Roman times, the
quince was picked
green, submerged in
honey, and left to
ripen in time to serve
at Roman wedding
feasts as a perfect sym-
bolic dessert.

❦

WEST MEETS EAST
Occidentalis. The
American planetree,
Platanus occidentalis, is
better known as the
sycamore. It can reach
a height of 150 feet.
Mike Dirr refers to it
as "a behemoth in the
world of trees."
Another striking fea-
ture is its exfoliating
bark, which exposes
the white inner bark.
Its Asian counterpart
is *Platanus orientalis.*

❦

SUMMER GARLANDS
Odoratus. Galium odoratum, sweet woodruff. In his *Herbal,* John Gerard notes that roses and sweet woodruff "being made up into garlands or bundles and hanged up in houses in the heat of Summer, doth very well attemper the air, cool and make fresh the place to the delight and comfort of such as are therein."

❦

FORGOTTEN GREENS
Oleraceus. Brassica oleracea var. *acephala,* or kale, is a vegetable that seems to have been all but forgotten. A member of the cabbage family, it was very popular among the Romans and Greeks, whose climates were unsuitable for growing cabbage. Full of vitamins and never bitter, kale is superb steamed.

❦

octan'drus: with eight anthers

octopet'alus: having eight petals

ocula'tus: having eyes; conspicuous

ocymoi'des: resembling basil, *Ocimum*

odessa'nus: of or from Odessa, southern Russia

odonti'tes: tooth-like; used for treating toothaches

odontochi'lus: with toothed lip

odoratis'simus: very fragrant; strong-smelling

odora'tus, odo'rus: fragrant; scented; sweet-smelling

officina'lis: medicinal; of the pharmacopoeia

officina'rum: of the apothecary's shop; medicinal

olb'ia: rich; of or from Hyères (formerly Olbia), France

oleafo'lius, oleifo'lius: with leaves resembling the olive, *Olea*

oleif'era: producing oil

oleoi'des: resembling the olive, *Olea* →

olera'ceus: used as food; cultivated

oli'dus: stinking; smelly

oligan'thus: producing few flowers

oligocar'pus: producing few fruits

olito'rius: pertaining to vegetable gardens; used as food

oliva'ceus: resembling the olive, *Olea*

olym'picus: of Olympus, mountain home of the gods; Olympus, the site of the Olympian games

onopor'dum: literally "ass-fart," refers
to the effect Scotch thistle, *Onopordum*, has
on donkeys who consume it

opa'cus: shaded; dark; dull

opercula'tus: with a lid or cover

ophiocar'pus: producing snake-like fruits

ophioglossifo'lius: with leaves like the
adder's-tongue fern, *Ophioglossum* →

opori'nus: of the late summer; autumnal

oppositiflo'rus: with opposite flowers

orar'ius: growing along the shoreline

orbicula'ris, orbicula'tus: round; like a wheel

orcaden'sis: from the Orkney Islands, Scotland

orchid'eus: resembling the orchids

orchidiflo'rus: producing orchid-like flowers

orchioi'des, orchio'des: resembling the orchids

orega'nus: of or from Oregon, United States

oreoph'ilus: mountain-loving

ores'bius: growing on the mountain

orgya'lis: about six feet in length; the distance
from fingertip to fingertip when the arms
are extended

orienta'lis: eastern; of the dawn

origanoi'des: resembling *Origanum*
(oregano or marjoram)

DARK AND HARDY
Opacus. Ilex opaca,
American holly. This
evergreen holly can
survive a wide range of
soil, light, and climatic
conditions. Old-time
weather forecasters
always checked the
number of berries to
predict the harshness
of the coming winter.
Few berries meant a
mild winter—birds
would be able to find
food in other places; a
lot of berries foretold
a hard winter—the
birds would need extra
nourishment.

❧

SHAKESPEARE'S SPICE
Origanoides. Origanum is
a genus that includes
the herbs oregano and
marjoram, inspiration
of poets as well as
cooks. Of it Shake-
speare wrote: "Indeed,
sir, she was the sweet
Marjoram of the
Salad, or rather the
Herb-of-grace."

A LEAF FOR LUCK
Ornans. Fraxinus ornus,
flowering ash. An ash
leaf with an equal
number of divisions on
each side represents
good luck. You should
pick the leaf and say
this English rhyme:
"Even ash, I do thee
pluck, ∾ Hoping thus
to meet good luck. ∾
If no good luck I get
from thee, ∾ I shall
wish thee on the tree."

❧

UNWELCOME GUEST
*Oxyacanthus. Crataegus
oxyacanthus,* hawthorn.
Hawthorn, an excellent
plant for strong
(though spiky) hedges,
was used in medieval
England to decorate
front doors on May
Day, but to bring it
into the house was said
to invite death inside
as well.

❧

or'nans, ornatis'simus: showy; splendid; adorned

orna'tus: adorned; embellished; handsome

ornithoceph'alus: shaped like a bird's head

ornitho'podus: shaped like a bird's foot

ornithorhyn'chus: shaped like a bird's beak

ortho'botrys: producing upright clusters

orthocar'pus: producing upright fruit

orthop'terus: with upright wings

osman'thus: fragrant-flowered

ossifra'gus: used for healing broken bones

ostrea'tus: as if covered with oyster shells

ostruthi'us: purplish in color

ouletrich'us: like curly hair

ovalifo'lius: with oval leaves

ova'lis: oval; egg-shaped

ova'tus: ovate

ovif'era, ovig'era: producing eggs
or egg-shaped features

ovi'nus: pertaining to sheep

oxyacan'thus: with sharp thorns →

oxygo'nus: with acute angles

oxyphi'lus: growing in acid soils

oxyphyl'lus: with sharp leaves

abula'rius: used for fodder or pasturage; having to do with foraging

pachyan'thus: flowering thickly; having flowers whose petals are thick

pachyphloe'us: having thick bark

pacif'icus: of the Pacific Ocean, usually of the North American Pacific coast

pagan'us: of country areas; rustic; of the wild

palaesti'nus: of or from Palestine

palea'ceus: with chaffy bracts; chaffy; scale-like

pal'lens: pale; fading; sallow

pallia'tus: cloaked; resembling a hooded, Greek-style cloak

pallidispi'nus: with pale spines

pal'lidus: pale; pale green in color ↗

pallifla'vens: pale yellow

palma'ris: palmate; of a hand's width; excellent; deserving the victor's palm

palma'tus: shaped like palm leaves; with five or more veins arising from a point

palmen'sis: of or from Las Palmas, Canary Islands

paludo'sus, palus'tris: marsh-loving; found in bogs

panicula'tus: having flowers in a cluster, with each flower borne on a separate stalk

THIN-SKINNED POETS
Pachyanthus. The prefix *pachy-*, from the Greek word meaning thick or massive, can describe abundance, but generally refers to thickness—of bark, flowers, or skin. The poet John Keats wrote that "A man cannot have a sensuous nature and be pachydermatous at the same time."

TURNIP PLASTER
Pectoralis. Before modern medicines were developed, a "pectoral" was often prescribed for illnesses of the chest and lungs. The Earl of Chesterfield, in 1749, recommended to his son pectorals of sage, barley, and turnips. In 1830 botanist John Lindley reported that licorice roots "contain a sweet subacid mucilaginous juice, which is much esteemed as a pectoral." He also noted that the leaves of ferns "generally contain a thick astringent mucilage, with a little aroma, on which account many are considered pectoral and lenitive."

pannon'icus: of or from Hungary, formerly Pannonia

panno'sus: ragged; shabby; shriveled

panormita'nus: of or from Palermo, Italy

papavera'ceus: resembling the poppy, *Papaver*

papilliona'ceus: resembling a butterfly →

papillo'sus: with papillae; having minute nipple-like protuberances on the surface

papyra'ceus: papery; with the texture of paper; resembling the Egyptian sedge, *Cyperus papyrus*

papyrif'era: used for producing paper; with paper-like bark

paradisi'acus: of parks or gardens; cultivated

paradox'us: strange; unexpected

parali'as: growing at the seaside

parasit'icus: parasitic; living off other plants

pardalian'thes: with the power to strangle a leopard

pardi'nus: with leopard-like spots

parietar'ius: growing on or near a wall

parmula'tus: resembling a small, round Roman shield

parnassi'acus: of or from Mount Parnassus, Greece

parti'tus: parted; deeply divided

par'vulus: very small

par'vus: small

pastina'ca: used as food; edible; resembling a pan or dish

patagon'icus: of or from Patagonia, tip of South America

patavi'nus: of or from Padua, Italy

patella'ris: circular; like a small dish or pan

pa'tens: spreading out from the stem; spreading extensively

pat'ulus: spreading; opened up; broad

pauciflo'rus: producing few flowers

pauc'us: few; not many; of the select few

pauper'culus: poor; meager

pavoni'nus: peacock blue in color

pectina'tus: comb-like; used for carding wool

pectora'lis: shaped like a breastbone; for treating chest colds

pecua'rius: of sheep or cattle; of the pasture

peda'lis: about a foot in length or height

pedatif'idus: cut or incised so that it resembles a foot

peda'tus: shaped like a human or bird's foot

pedemonta'nus: of or from the Piedmont, Italy

pedicula'rius: louse; lousy; resembling lice

peduncula'ris, peduncula'tus: bearing a peduncle (a flower cluster supported by a single stalk) →

pedunculo'sus: with many peduncles

pellu'cidus: transparent; clear; shining; pleasing

pelta'tus: resembling a small, leather shield used in the Roman army; having leaves with stems attached to the inner surface of the leaves

pelvifor'mis: shaped like a basin or shallow cup

penduliflo'rus: producing flowers that hang

pen'dulus: hanging down; drooping

DAINTY BIRDS' FEET
Pedatus. The bird's foot violet, *"Viola pedata* and its form, *bicolor,* are exquisite natives. The problem is finding exactly the right location for them. They are often found growing along roadsides. . . . Some authorities . . . recommend growing them in acid sand . . . others suggest clay . . . the critical factor is good drainage." Elizabeth Lawrence, *A Rock Garden in the South,* 1990.

SWEET DREAMS
Peregrinus is a word derived from the Latin *peregri*, meaning away from home. Applied to plants *peregrinus* may mean not native or that it spreads, as in the case of the native sweet fern, *Comptonia peregrina*. American colonists used its fragrant leaves to stuff their mattresses—assuring themselves sweet dreams.

PERSIAN APPLE ↑
Persicaefolius, persicifolius. The peach, *Prunus persica*, was originally called *Malum persicum*, the "Persian apple."

penicilla'tus: covered with tufts of hair; resembling the bristles of a brush

peninsula'ris: peninsular

penna'tus: feathered; winged

pennig'era: bearing feather-like structures

penniner'vis: with veins arranged like a feather

pennsylvan'icus: of or from Pennsylvania, United States

pen'silis: hanging; suspended from above

pentade'nius: five-toothed

pentago'nus: with five angles

pentapetaloi'des: like or appearing to be five petals

pe'po: sun-ripened

perbel'lus: very beautiful

percus'sus: sharp-pointed; actually or just appearing to be perforated

peregri'nus: exotic; foreign; strange

peren'nans, peren'nis: continuing through the years; living more than two years

perfolia'tus: with leaf surrounding the stem so that the stem appears to pass through the leaf

perfora'tus: pierced with holes

pergrac'ilis: very slender

permix'tus: much mixed; confused

perpropin'quus: closely related

persicaefo'lius, persicifo'lius: with leaves resembling the peach, formerly *Persica vulgaris*

per'sicus: of or from Persia; resembling the peach

perspic'uus: clear; transparent; evident

pertu'sus: perforated; tattered

perula'tus: with conspicuous scales, as on a bud

peruvia'nus: Peruvian

petaloid'eus: resembling a petal

petiola'ris, petiola'tus: petioled, that is, having a leaf stalk

petrae'us: growing among the rocks

petrocal'lis: rock beauty

phaeocar'pus: producing dark fruits

phae'us: dusky; dark; brown in color

philadel'phicus: of or from Philadelphia, Pennsylvania, United States

phleban'thus: with conspicuously veined flowers

phleioi'des: resembling the pasture and hay grass, *Phleum*

phlogiflo'rus: flame-flowered; phlox-flowered †

phlogifo'lius: with leaves like *Phlox*

phoenic'eus: purple-red in color; of or from Phoenicia

phoenicola'sius: with purple hair

phryg'ius: of or from Phrygia, Asia Minor; Trojan

phu': rotten-smelling; acrid; having a foul odor (from the Greek for valerian)

THE FIREBIRD'S TREE *Phoenicolasius.* Roman naturalist Gaius Plinius Secundus (Pliny the Elder) suggests that the name of the date tree, genus *Phoenix,* and the mythical firebird are connected: "For it was assured to me that the said bird died with that tree, and revived of it selfe as the tree sprung againe." Ephraim Chambers, in his *Cyclopaedia,* wrote that "the Phoenicians gave the name 'phoenix' to the palm-tree by reason when burnt down to the very root, it rises again fairer than ever." The *Oxford English Dictionary* notes that "some would explain . . . the date as 'the red fruit.'" The Phoenicians may have been so named because they came from red lands to the east.

MYTHICAL GROVES ↑
Pineus. Loblolly pine, Georgia pine, yellow jack, pitch pine, torch pine, light-wood, white pine, old field pine, longshucks, bastard, foxtail, Indian, and rosemary pine. These common names given to the pine trees of the New World celebrate the industry and experience of the colonists who found them here and are far removed from the mythological grandeur of the pine groves once sacred to the goddess Diana, the Huntress.

phyllanthoi'des: resembling the Otaheite gooseberry tree, *Phyllanthus acidus*

phyllomani'acus: excessively leafy; producing a riot of foliage

phymatochi'lus: long-lipped

phytolaccoi'des: resembling pokeweed, *Phytolacca americana*

pice'us: dark as pitch; black in color

pictura'tus: as if painted upon; variegated; embellished

pic'tus: painted; brightly marked

pilea'tus: resembling a cap or skullcap

pilif'era: bearing short, soft hairs

pilo'sus: shaggy; with long, soft hairs

pilula'ris: resembling a ball or globe

pimpinellifo'lius: with leaves like anise, *Pimpinella anisum*

pineto'rum: of pine forests

pin'eus: resembling the pines, *Pinus;* cone-bearing

pinguifo'lius: thick-leaved; with oily leaves

pinifo'lius: with leaves like the pine

pinnatifo'lius: with feather-shaped leaves

pinna'tus: resembling a feather in structure; having similar parts arranged oppositely on a central axis

piperi'ta: resembling peppermint; sharply fragrant or flavored

pisif'era: producing peas or pea-like seeds

placa'tus: quiet; calm; gentle

placentifor'mis: ring-shaped; shaped like a cake

plan'ipes: flat-footed; with a flat stalk

plantagin'eus: resembling the plantain, *Plantago* ↓

pla'nus: of one plane; flat

platantoi'des: resembling the plane tree, *Platanus*

platycan'thus: with broad, flat spines

platycau'lon: with broad, flat stems

platy'podus, plat'ypus: with a broad foot; with a wide stalk

pleioneu'rus: marked by many veins

pleniflo'rus: with abundant or double flowers

ple'nus: full; double

pleuros'tachys: side-spiked

plica'tus: pleated; folded lengthwise

pluma'rius, pluma'tus: plumed; covered with feathers

plumbag'inoi'des: resembling *Plumbago*; lead blue in color

plum'beus: the color of lead

plumo'sus: feathery; downy

pluvia'lis: growing in rainy places

pocopho'rus: bearing fleece; covered in fleece

poculifor'mis: shaped like a deep drinking cup

podag'ricus: used to treat gout or arthritis

podocar'pus: bearing fruits on a stalk; literally "foot-fruit"

podol'icus: of Podolia, southwestern Russia

podophyl'lus: with stalked leaves

poet'icus: pertaining to poets

PEAS PLEASE

Pisifera. Chamaecyparis pisifera, a huge tree, was given its specific name because of the shape of its female cones. Most gardeners are more familiar with the smaller cultivars such as 'Filifera Aurea,' with golden thread-like branchlets, and 'Globosa,' a slow-growing rounded form.

❀

TRUE BLUE

Plumbaginoides. Ceratostigma plumbaginoides, blue leadwort, gets its common and specific names from the color of its flowers. Summer-blooming and deciduous, it makes an ideal groundcover for sunny areas. Underplant it with small bulbs, which will flower while the leadwort is dormant. As it leafs out the leadwort will disguise the bulbs' dying foliage.

TRANQUILIZER

Ponticus. Daphne pontica is an evergreen spreading shrub whose yellow-green flowers release a sharp, spicy fragrance in evening. British garden writer Christopher Lloyd recommends planting it somewhere along the path between house and car, commenting, "It'll take your mind off your problems."

❧

POTENT POTATOES

Potatorum. The word "potable," or drinkable, has nothing to do with potatoes, except where vodka is concerned. In regard to its potable side products, Ruskin said, "In the potato, we have the scarcely innocent underground stem of one of a tribe set aside for evil."

❧

pola'ris: of or from the North Pole →

polifo'lius: with white or gray leaves; with leaves resembling germander, *Teucrium*

poli'tus: polished; elegant

pollica'ris: the length of the last thumb joint

polo'nicus: of or from Poland

polyan'drus: with many stamens

polybot'rya: producing many grape-like clusters

polybul'bon: with many bulbs

polyg'amus: having various combinations of sexual structures on one plant, or on separate plants of the same species

polyg'yrus: twining

poly'lepis: covered with many scales

polymor'phus: of many forms; variable in shape

polysper'mus: producing many seeds

polystic'tus: marked by many dots

poma'ceus: resembling the apple, *Malus*

pomeridia'nus: of the afternoon; flowering after noon

pomif'era: producing apples or apple-like fruits

pomos'us: loaded with fruit

pondero'sus: heavy; large; of great weight

ponti'cus: from Pontus, the southern end of the Black Sea

populifo'lius: with leaves like the poplar, *Populus*

porci'nus: pertaining to swine

porophi'lus: growing on stony ground

porphy'reus: purple in color

porrifol'ius: with leaves like the leek, *Allium porrum*

por'rigens: spreading

potamoph'ilus: river-loving; growing in the rivers

potato'rum: having to do with drinkers; used in fermentation

praeal'tus: very tall; showy

prae'cox: developing earlier than most of the rest of its genus; premature

praemor'sus: appearing to have been nibbled on

praerup'torum: growing in rough places

prae'stans: distinguished; excelling

praetex'tus: bordered; edged; fringed

pras'inus: grass green in color

praten'sis: growing in meadows

prav'us: crooked; deformed; irregular

precato'rius: relating to prayers

pren'ans: drooping

primuloi'des: resembling primrose, *Primula* ↗

prin'ceps: princely; first in rank; distinguished

prismat'icus: resembling a prism

proboscid'eus: with a proboscis; resembling a long flexible snout

PAINTED WITH DELIGHT
Pratensis. Cardamine pratensis, lady's-smock. If the name leaves any doubt, Shakespeare confirms lady's-smock as a meadow-growing species with these lines: "Daisies pied, and Violets blue, ∾ And Lady-smocks, all silver white, ∾ And Cuckoo-birds of yellow hue ∾ Do paint the meadows with delight."

⚘

PATH OF DALLIANCE
Primuloides. Primrose was the flower of imagination for many centuries. It was believed to be magical, healthful, and useful as a beauty treatment. Shakespeare warned beware he who "the primrose path of dalliance treads."

⚘

A TASTE FOR TEA
Procumbens. The botanical name for the creeping wintergreen shrub, or teaberry, is *Gaultheria procumbens.* Alice Morse Earle notes in *Old Time Gardens* that mountain children liked to carry teaberry leaves to church. They chewed them for their wintergreen taste during the long sermons.

proce'rus: tall or long

procum'bens: prostrate; flat on the ground

procur'rens: spreading outward; moving ahead

prodig'iosus: wonderful; marvelous; enormous; prodigious; monstrous; pertaining to omens

produc'tus: lengthened; stretched out

prolif'era: reproducing by offshoots or young plantlets

prolif'icus: fruitful; abundant

propen'dens: hanging down

propin'quus: related; closely allied to; neighboring

prostra'tus: prostrate; lying flat without rooting

protru'sus: protruding; projecting beyond its boundaries

provincia'lis: provincial; rustic; of or from Provence, France

pruina'tus, pruino'sus: with a hoary bloom; appearing to be frosted over

prunelloi'des: resembling the herb *Prunella*

prunifo'lius: with leaves like *Prunus* (cherry, plumb, apricot) ←

pru'riens: itching; causing irritation; stinging

psiloste'mon: with slender or naked stamens

psittac'inus: parrot-like in coloration

psyco'des: fragrant; resembling butterflies

psyl'lius: flea-like, mostly referring to seeds

ptarmi'cus: causing sneezing

pu'bens: downy; with soft hair; juicy; mature

pubig'era: down-bearing; covered in down

pudi'cus: bashful; retiring; virtuous; pure

pugionifor'mis: shaped like a dagger

pulchel'lus: pretty; beautiful

pul'cher: handsome; beautiful

pul'lus: dark-colored; raven black; mournful

pulverulen'tus: powdered; dust-covered

pulvina'tus: cushion-like in shape

pu'milus: dwarf; low-growing; small

puncta'tus: pricked; dotted; with a pocked surface

pun'gens: piercing; sharp-pointed; pricking

pun'ica: of or from Carthage

punic'eus: crimson in color

pur'gans: purgative

purpura'ceus, purpura'tus, purpu'reus: purple in color

pusil'lus: very small; insignificant; petty

pustula'tus: covered with blisters or pimples

puteo'rum: of wells; pits; dungeons

pycnacan'thus: densely spined

pycnan'thus: densely flowered

pygmae'us: pygmy; very small for its kind

pyramida'lis: pyramidal in shape

pyrenae'us, pyrena'icus: of the Pyrenees, France and Spain

pyrifor'mis: shaped like a pear, *Pyrus*

pyxida'tus: box-like; with a lid

APPLES OF CARTHAGE ↑
Punica. The pomegranate, originally known to the Romans as the "apple from Carthage" (*Malus punica*), later became known as the "apple full of grains" (*Pomum granatum*). Now the modern botanical name, *Punica granatum*, incorporates both elements.

FALL SPLENDOR
Quercifolius. Hydrangea quercifolia. Eighteenth-century American botanist and plant collector John Bartram discovered a southeastern hydrangea with coarse leaves like those of the red oak. He named it *Hydrangea quercifolia.* Foot-long panicles of white flowers produced in May and June fade to apricot, rose, and then to brown. Even more remarkable is its autumn foliage colors—red, orange, and burgundy. And in winter, when the leaves have dropped, a bright, peeling cinnamon-colored bark is revealed.

✤

quadrangula'ris, quadrangula'tus: with four angles

quadra'tus: in four or fours; divided into four parts; producing four-part structures

quadriauri'tus: having four ears; four-lobed

quadri'color: four colors

quadridenta'tus: with four teeth

quadrif'idus: cut into four parts

quadrifo'lius: four-leaved

quadriparti'tus: parted in four ways

quadrival'vis: with four valves

quadrivul'nerus: exhibiting four wounds

quercifo'lius: with leaves resembling oak, *Quercus*

querc'inus: of or resembling oak, *Quercus*

quina'tus: in fives; divided into five parts

quinque'color: exhibiting five colors

quinqueflo'rus: with five flowers ↗

quinquefo'lius: with five leaves

quinquelocula'ris: five-celled

quinquener'vis: with five conspicuous nerves or veins

quinquepuncta'tus: five-spotted

quinquevul'nerus: five wounds or marks

quisquilico'lus: living on or in rubbish

acem'ifer: producing clusters; appearing to be covered with racemes

racemo'sus: flowers growing in racemes (flowers opening along a central stalk or axis from the bottom up, like the lily-of-the-valley)

ra'dians: radiating outward; spoke-like; shining; beautiful

radi'cans: with structures that have rooting abilities

radica'tus: having or marked by conspicuous roots

radico'sus: many-rooted

radi'cum: referring to roots

radio'sus: with many rays

rad'ula: rough; like a scraper

ragusi'nus: of or from Dubrovnik (formerly Ragusa), Yugoslavia

ramenta'ceus: covered with fine down

ramiflo'rus: with branching inflorescences

ramondioi'des: resembling the low-growing European mountain herb *Ramonda*

ramo'sus: branched; twiggy; twig-like

ranunculoi'des: resembling the buttercup, *Ranunculus*, in appearance or habitat

rapa'ceus: of or about turnips; often tiny seedlings that resemble newly sprouted turnips

BUTTER LOVERS ↓
Ranunculoides. An old custom says that if you hold a buttercup under a person's chin and the skin reflects its yellow color, he loves butter.

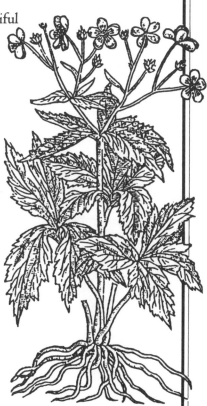

COFFIN WOOD
Recurvus. Juniperus recurva
is a large evergreen
shrub with curved
branches and drooping
branchlets. Native to
the Himalaya Moun-
tains, *Juniperus recurva*
var. *coxii*, the coffin
juniper, has very
resinous wood that is
used for incense in
Buddhist temples and
for coffins.

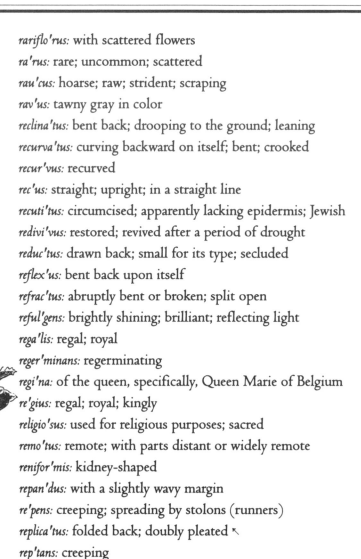

rariflo'rus: with scattered flowers

ra'rus: rare; uncommon; scattered

rau'cus: hoarse; raw; strident; scraping

rav'us: tawny gray in color

reclina'tus: bent back; drooping to the ground; leaning

recurva'tus: curving backward on itself; bent; crooked

recur'vus: recurved

rec'us: straight; upright; in a straight line

recuti'tus: circumcised; apparently lacking epidermis; Jewish

redivi'vus: restored; revived after a period of drought

reduc'tus: drawn back; small for its type; secluded

reflex'us: bent back upon itself

refrac'tus: abruptly bent or broken; split open

reful'gens: brightly shining; brilliant; reflecting light

rega'lis: regal; royal

reger'minans: regerminating

regi'na: of the queen, specifically, Queen Marie of Belgium

re'gius: regal; royal; kingly

religio'sus: used for religious purposes; sacred

remo'tus: remote; with parts distant or widely remote

renifor'mis: kidney-shaped

repan'dus: with a slightly wavy margin

re'pens: creeping; spreading by stolons (runners)

replica'tus: folded back; doubly pleated ↖

rep'tans: creeping

resec'tus: cut off

resinif'era: resin-bearing

resino'sus: full of resin

reticula'tus: reticulate; netted

retino'des: retained

retor'tus: twisted back

retroflex'us: reflexed

retrofrac'tus: broken or bent backward

retu'sus: retuse, that is, notched slightly at a rounded apex

rever'sus: reversed.

revolu'tus: revolute; rolled backward

rex': king

rhamnoi'des: resembling the buckthorn tree, *Rhamnus*

rhexifo'lius: with leaves resembling the native meadow beauty, *Rhexia*

rhipsalioi'des: resembling the cactus genus, *Rhipsalis*

rhododen'sis: of or from the Isle of Rhodes, Greece

rhomboid'eus: diamond-shaped

rhytidophyl'lus: wrinkle-leaved

ricinoi'des: resembling the ornamental castor bean, *Ricinus* →

ri'gens: rigid; stiff

rin'gens: gaping

ripa'rius: growing along riverbanks

riva'lis: pertaining to brooks

rivula'ris: brook-loving

SOUNDALIKES

Resinifera, resinosus. The words resin and rosin are commonly confused. Resin refers to any number and variety of substances of plant origin. Rosin has a much more specific meaning. It is the dark, solid substance made from the resin of pine trees and used on the bows of stringed musical instruments to increase friction.

GARLIC AND ROSES
Rosaceus. Planting a clove or two of garlic beside rosebushes repels aphids and won't affect the flowers' delicate perfume. "A rose is a rose is a rose."
Gertrude Stein

❦

DEATH'S SOFT PETALS
Rotundifolius. Drosera rotundifolia, round-leaved sundew, is an insectivorous plant whose leaves, covered with sticky hairs, are said to sparkle in the sun as though covered with dewdrops. Swinburne wrote: "You call it sundew; how it grows, ∾ If with its color it have breath, ∾ If life taste sweet to it, if death ∾ Pain its soft petal, no man knows: ∾ Man has no sight or sense that saith."

❦

robus'tus: robust; stout

roma'nus: Roman

rosa'ceus: rose-like; covered with roses

rosmarinifo'lius: with leaves like the rosemary, *Rosmarinus* ↓

rostra'tus: rostrate; beaked as a ship; decorated with such flowers as are appropriate for a military celebration

rosula'ris: in rosettes

rota'tus: wheel-shaped

rotundifo'lius: round-leaved

rubelli'nus, rubel'lus: reddish

ru'bens, ru'ber: red; ruddy

rubicun'dus: rubicund; red

rubigino'sus: rusty

rubioi'des: resembling madder, *Rubia tinctorum*

rubricau'lis: red-stemmed

rubroner'vis: red-veined

ru'dis: wild; not tilled; not grown on cultivated land

rudius-'culus: wild; untamed

rufes'cens: almost red; light red; tawny

rufiner'vis: with red nerves or veins

ru'fus: red; reddish

rugo'sus: rugose; wrinkled

runcina'tus: runcinate, that is, with the leaves cut to the leaf base as in a dandelion

rupes'tris: of or pertaining to cliffs

rupic'olus: growing in the cliffs

rupif'ragus: growing on cliffs

ruscifo'lius: with leaves like the butcher's-broom, *Ruscus*

russa'tus: reddish; russet

rustica'nus, rus'ticus: rustic; pertaining to the country

ruthen'icus: from Ruthenia, now known as western Russia

rutidobul'bon: rough-bulbed

rutifo'lius: with foliage like that of rue, *Ruta;* with blue-gray foliage ↘

ru'tilans: red; becoming red

RUSTIC RADISHES
Rusticanus. Armoracia rusticana—horseradish—is a traditional or "rustic" condiment in eastern Europe, where it grows wild. It is cultivated for its root, whose white flesh has a sharp, hot taste. To make a cold horseradish sauce: Soak breadcrumbs in cream and squeeze them dry. Add grated horseradish, sugar, salt, heavy cream, and vinegar. Adjust to taste.

HERB OF GRACE
Rutifolius. "There's fennel for you, and columbines; there's rue for you; and here's some for me; we may call it herb of grace o'Sundays. O! you must wear your rue with a difference." William Shakespeare, *Hamlet,* 1602

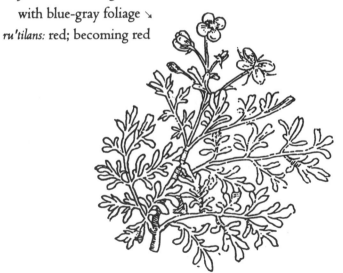

SWEET AND PRETTY
Saccharum. One of autumn's most spectacular foliage displays is provided by the sugar maple, *Acer saccharum.* Its sap is the source of maple sugar and maple syrup. Milled, its trunk provides lovely golden-colored maple wood for furniture and cabinetry.

❀

OUT, DAMNED SPOT
Sanguineus. Genus names can be as descriptive as species names; to wit, *Sanguinaria canadensis,* or bloodroot. "Care should be taken in picking the flower . . . as the red liquid which oozes blood-like from the wounded stem makes a lasting stain. It was prized by the Indians as a decoration for their faces."
Mrs. William Starr Dana, *How to Know the Wild Flowers,* 1990

acchara'tus: containing sugar; sweet

saccharif'era: sugar-bearing ⟍

sacchar'inus: saccharine

saccharoi'des: like sugar

sac'charum: of sugar

saccif'era: having a sack-like swelling, said of petals, sepals, and sometimes of stamens and leaves

sacro'rum: sacred; of or from sacred places

sagitta'lis, sagitta'tus: sagittate; resembling an arrowhead

sagittifo'lius: with arrowhead-shaped leaves

salicariaefo'lius: with leaves like the willow, *Salix,* often long and narrow

salicifo'lius: with leaves like the willow, *Salix*

salic'inus: resembling the willow, *Salix*

salicornioi'des: resembling *Salicornia,* glasswort, an herb with horn-like branches that thrives in salt marshes

salig'nus: of the willow

sali'nus: salty; growing in salty places

salsugino'sus: of the salt marsh

salviaefo'lius, salvifo'lius: *Salvia*-leaved

sambucifo'lius: with leaves resembling those of the elder, *Sambucus*

sambuci'nus: resembling *Sambucus*

sanc'tus: holy

sanguin'eus: bloody; blood red

sap'idus: savory; pleasing to taste

sapien'tum: wise; pertaining to knowledge

sapona'ceus: soapy; forming lather

sarco'des: fleshy; resembling flesh

sarmat'icus: of Sarmatia, south Russia; Russian

sarmento'sus: having runners or stolons

sati'vus: cultivated; planted deliberately

satura'tus: saturated; filled with moisture

sauroceph'alus: lizard-headed

saxat'ilis: found among rocks

saxic'olus: growing among rocks

saxo'sus: full of rocks

sca'ber: scabrous; rough

scaber'rimus: very rough

scabiosaefo'lius: with leaves resembling the mourning bride,
   Scabiosa

scabrel'lus, scab'ridus: somewhat rough

scan'dens: scandent; climbing or growing upward

scapo'sus: with scapes (leafless flower stalks)

scario'sus: scarious; shriveled; thin, dry; often translucent
   and not green

scep'trum: of a scepter

SOAP BY THE BAR ↑
*Saponaceus. Saponaria
officinalis,* called bounc-
ing bet, contains in its
roots and leaves
saponin, which lathers
in water. The common
name probably came
from the use of the
branches to clean beer
bottles by barmaids
who were called Bet or
Betsy in early England.

❦

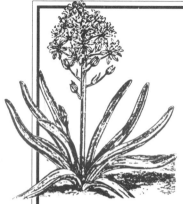

GHOST FLOWERS ↑
Scilloides. Precocious
bloomers, *Puschkinia
scilloides* may flower as
early as the first week
of January and often as
not, flowers appear at
ground level on stems
barely pushed out of
the earth. A bit of
warmth on a sunny
winter day releases the
spicy fragrance. Its
fresh green foliage is
similar to the scillas,
which this little bulb
resembles. Of this
resemblance E. A.
Bowles wrote "a pretty
little grey thing like
the ghost of a Scilla
come back to earth."

schidig'era: bearing spines

schisto'sus: schistose; easily split or divided

schizoneu'rus: having divided nerves

schizopet'alus: with cut or divided petals

schizophyl'lus: with divided leaves

schola'ris: pertaining to a school

scilloi'des: resembling the squill (formerly *Scilla,* now
   Hyacinthoides)

sclerocar'pus: having hard fruits

sclerophyl'lus: having hard leaves

scopa'rius: resembling a broom (from the Latin for
   "floor-sweeper")

scopulo'rum: of rocks

scorpioi'des: resembling a scorpion

scorzoneroi'des: resembling black salsify, *Scorzonera*

scot'ica: Scotch

sculp'tus: carved or sculpted

scuta'tus: shaped like a buckler (a small,
   round shield)

scutella'ris, scutella'tus: salver- or dish-shaped

scu'tum: like a small, round shield

sebif'era: tallow-bearing

sebo'sus: full of tallow or grease

sechella'rum: of or from the Seychelles

seclu'sus: hidden; secluded; isolated

secundiflo'rus: having flowers on only one side of the stem

secun'dus, secunda'tus: one-sided; secund; having leaves or flowers on only one side of a stem

securig'era: axe-bearing, refers to the shape of the seed pods

seg'etum: of cornfields

selaginoi'des: Selago-like; clubmoss-like

semiala'tus: semi-winged; half-winged; imperfectly winged

semibacca'tus: semi-berried

semicauda'tus: semi-tailed

semicylin'dricus: semi-cylindrical

semidecan'drus: half ten-stamened

semipinna'tus: imperfectly pinnate

semperflo'rens: ever-flowering

semper'virens: evergreen

sempervivoi'des: resembling the house leek, *Sempervivum*

senecioi'des: resembling *Senecio*

seni'lis: senile; old; white-haired

sensib'ilis: sensitive

sensiti'vus: sensitive

sepia'rius: of or pertaining to hedges

se'pium: of hedges or fences

septangula'ris: seven-angled

septem'fidus: seven-cut

septem'lobus: seven-lobed

septempuncta'tus: seven-spotted

OLD AND TRUE ↑
Sempervirens. Sequoia sempervirens is the magnificent redwood of the coastal forests of Oregon and northern California, an evergreen tree that can grow up to 300 feet, and of which some trees are said to live to be 3,000 years old. Perhaps the only tree more massive than the redwood is its cousin, *Sequoiadendron giganteum.*

PERSISTENT CREEPERS
Serpens. The Latin *serpens, serpentis* plays upon the metaphorical equivalence between the serpentine path of the snake and the creeping growth of low-lying plants. Louise Beebe Wilder notes that common names such as "Meg-many-feet, Gill-over-the ground, Robin-run-the-hedge, Roaming Charlie, Creeping Jenny, Jack-jump-about, or Mother-of-thousands" commemorate the "insistent colonizing proclivities" of creepers, whose persistent growth knows no boundaries.
Louise Beebe Wilder, *Color in My Garden*, 1990

septentriona'lis: northern

sepul'tus: sepulchered; interred; buried

serican'thus: silky-flowered

seric'eus: silky; having soft, silky hairs

sericif'era, sericof'era: silk-bearing

serot'inus: late; late-flowering; late-ripening

ser'pens: creeping; crawling

serpenti'nus: of snakes; serpentine

serpyllifo'lius: thyme-leaved; *Serpyllum*-leaved

serratifo'lius: serrate-leaved

serra'tus: serrate; saw-toothed ↘

serrula'tus: somewhat serrate

sesquipeda'lis: one and one-half feet high or long

sessiflo'rus: having stalkless flowers

sessifo'lius: having stalkless leaves

sessiliflo'rus: with stalkless flowers

sessilifo'lius: with stalkless leaves

ses'silis: sessile; stalkless

seta'ceus: bristle-like; resembling bristles

setifo'lius: bristle-leaved; with leaves resembling bristles

setig'era, set'iger: bristle-bearing

seti'podus: bristle-footed

setispi'nus: bristle-spined

seto'sus: full of bristles

setulo'sus: full of small bristles

sexangula'ris: six-angled

sia'meus: of Siam (now Thailand)

sibir'icus: of Siberia

siculifor'mis: dagger-formed

sic'ulus: of or from Sicily →

siderophloi'us: iron bark

siderox'ylon: iron wood

signa'tus: marked; designated

silaifo'lius: with leaves like those of celery (from the Latin word *silaus,* meaning a plant resembling celery)

silic'eus: pertaining to or growing in sand

siliculo'sus: bearing silicles, two carpelled fruits that are twice as long as they are wide

siliquo'sus: bearing siliques, two carpelled fruits that are three or more times longer than wide

silvat'icus, silves'tris: pertaining to woods

sim'ilis: similar; like

sim'plex: simple; unbranched

simplicicau'lis: simple-stemmed

simplicifo'lius: simple-leaved

simplicis'simus: simplest

sim'ulans: similar to; resembling

BRISTLY LEAVES
Setigera. Viburnum setigerum, the graceful tea viburnum, was discovered by plant collector E. H. "Chinese" Wilson. The species name describes the small bristles located at the ends of the leaf veins. Corymbs of white flowers in summer are followed by showy clusters of large fruits, which ripen to red color in autumn.

THE JUDAS TREE
Siliquosus. Cercis siliquastrum, love tree. This Eurasian tree with purplish rose flowers that blossom before the leaves emerge is also known as the Judas tree, from the belief that Judas Iscariot hanged himself from its branches.

BEDTIME STORY ↑
Somnifera. Papaver som-niferum, opium poppy.
Ceres, goddess of agriculture, became so exhausted searching for her lost daughter, she neglected the wheat. Somnus, god of sleep, created the opium poppy for her use. Once Ceres was properly rested, the wheat began to flourish once again. As Keats wrote in "Sonnet to Sleep," "O soothest Sleep! if so it please thee, close, ∾ In midst of this thine hymn, my willing eyes, ∾ Or wait the Amen ere thy poppy throws ∾ Around my bed its lulling charities."

sin'icus: Chinese

sinua'tus, sinuo'sus: sinuate; wavy-margined

siphilit'icus: syphilitic; formerly used in the treatment of syphilis as *Lobelia siphilitica*

sisala'nus: pertaining to sisal, a fiber obtained from an *Agave* native to Yucatán, Mexico

sisymbrifo'lius: with leaves resembling *Sisymbrium*

smarag'dinus: of emerald

smilac'inus: of *Smilax* (a genus of the lily family)

sobolif'era: bearing creeping rooting stems or roots

socia'lis: sociable; companionable

socotra'nus: of Socotra (island off Arabia)

sodome'um: of Sodom, a biblical city destroyed by fire

solandriflo'rus: with flowers like those of *Solandra*

sola'ris: of the sun

soldanelloi'des: resembling *Soldanella*, a genus of mountain-dwelling herbs related to the primrose

sol'idus: solid; dense

somnif'era: sleep-inducing

sonchifo'lius: with leaves like those of *Sonchus*, sow thistle

sorbifo'lius: with leaves like those of *Sorbus*, mountain ash

sor'didus: dirty

spadic'eus: with a spadix, a fleshy spike of tiny flowers

sparsiflo'rus: sparsely flowered

sparsifo'lius: sparsely leaved

spar'sus: sparse; few

spar'teus: pertaining to the Spanish broom, *Spartium*

spatha'ceus: with a spathe (a leaf or bract surrounding a
   flower cluster or spadix)

spathula'tus: spatulate; spoon-shaped

spathulifo'lius: spatulate-leaved

speciosis'simus: very showy

specio'sus: showy; good-looking

spectab'ilis: spectacular; remarkable; showy

spectan'drus: showy

spec'trum: an image; apparition

specula'tus: shining, as if with mirrors

sphacela'tus: dead; withered; diseased

sphaer'icus: spherical

sphaerocar'pus: with rounded fruit

sphaeroceph'alus: round-headed

sphaeroid'eus: sphere-like;
   round

sphaerostach'yus: with rounded
   spikes

spica'tus: spicate; with spikes

spicifor'mis: spike-shaped

spicig'era: spike-bearing →

spiculifo'lius: spicule-leaved;
   needle-leaved; spiny-leaved

A BUM RAP

Speciosus. Solidago speciosa,
showy goldenrod.
Many folk beliefs sur-
round the goldenrod:
that the flower marks
the spot where treasure
is buried, that planting
goldenrod by the front
door insures good for-
tune to a household,
that in the right hands,
goldenrod can be used
as a divining rod to
locate water deep
underground, and that
goldenrod causes
sneezing. The showy
golden flowers of gold-
enrod do not cause
people to sneeze, the
pollen is too heavy.
Ragweed, which
blooms at the same
time as goldenrod, has
the irritating wind-
borne pollen that
starts people sneezing.

PERMISSIBLE BLUSHES ↑
Stachyoides. "Lamb's ears
(formerly *Stachys lanata*,
and still sold under
that name, but techni-
cally now *S. olympica* or
sometimes *S. byzantina*)
is a plant with gray
foliage so soft and
downy that I'm always
tempted to rub it
against my cheek—
and sometimes do. (An
old common name for
it is Quaker rouge,
commemorating the
former practice of
Quaker women who

spina'rum: spiny

spines'cens: somewhat spiny

spinif'era: bearing spines

spinosis'simus: very spiny

spino'sus: full of spines

spinulif'era: bearing small spines

spinulo'sus: somewhat or weakly spiny

spira'lis: spiral

spirel'lus: with a little spiral

splen'dens: splendid

splendidis'simus: very splendid

splen'didus: splendid

spondioi'des: resembling *Spondias*, a tropical tree related
    to the cashew

spuma'rius: frothing

spu'rius: spurious; false

squa'lens, squal'idus: squalid; filthy

squama'tus: squamate; with small scale-like leaves or bracts

squamo'sus: full of scales

squarro'sus: spreading in all directions; rough; protruding

stachyoi'des: resembling *Stachys*; having flowering spikes

stamin'eus: bearing prominent stamens

stans': standing; erect; upright

stauracan'thus: with cross-shaped spines

stella'ris, stella'tus: stellate; starry

stellip'ilus: with stellate hairs; with radiating branches

stellula'tus: somewhat stellate; star-like

stenocar'pus: narrow-fruited ↘

stenoceph'alus: narrow-headed

stenog'ynus: with a narrow stigma

stenopet'alus: narrow-petaled

stenophyl'lus: narrow-leaved

stenop'terus: narrow-winged

stenostach'yus: narrow-spiked

ster'ilis: sterile; infertile; barren

stigmat'icus: marked; of stigmas

stigmo'sus: much marked; pertaining to stigmas

stipula'ceus, stipula'ris, stipula'tus: having stipules
　(basal appendages of the petiole)

stipulo'sus: having large stipules

stolonif'era: producing stolons or runners that take root

stramineofruc'tus: with straw-colored fruit

stramin'eus: straw-colored

strangula'tus: strangled; constricted

streptocar'pus: with twisted fruit

streptopet'alus: with twisted petals

streptophyl'lus: with twisted leaves

streptosep'alus: with twisted sepals

striat'ulus: faintly striped

stria'tus: striated; striped

used the plant to bring a permissible blush to their faces when more frankly cosmetic substances were shunned as frivolous.)"
Allen Lacy, *The Garden in Autumn,* 1990

❧

A TRAVELING TOAD
Stolonifera. Tricyrtis stolonifera, a member of the clan known by the unfortunate name "toad lilies," has much to offer the autumn garden. From underground stolons, stems rise carrying flower buds. In October a succession of white flowers spotted with a deep wine color open. Plant it beside a shaded path.

❧

BLUEBONNETS
Subcarnosus. The Texas bluebonnet, *Lupinus subcarnosus,* gets both its common and specific name from its flowers. The shapes of the flower parts resemble the parts of an old-fashioned gingham bonnet and the flowers are fleshy to the touch.

strictiflo'rus: stiff-flowered

stric'tus: strict; upright; erect ✓

strigillo'sus: somewhat strigose, that is, having somewhat flat bristles or scales

strigo'sus: strigose

strigulo'sus: with small or weak appressed hairs

striola'tus: faintly striped

strobila'ceus: resembling a cone

strobilif'era: cone-bearing

struma'rius: of tumors or ulcers

struma'tus: with tumors or ulcers

strumo'sus: having cushion-like swellings

stylo'sus: with prominent styles

styphelioi'des: resembling *Styphelia,* an Australian shrub

styracif'luus: flowing with storax or gum

suave'olens: sweet-scented

sua'vis: sweet; agreeable

suavis'simus: sweetest

subacau'lis: without much of a stem; somewhat stemmed

subalpi'nus: nearly alpine

subauricula'tus: somewhat eared

subcaeru'leus: slightly blue

subca'nus: somewhat hoary or grayish white

subcarno'sus: rather fleshy; thick; soft

subcorda'tus: somewhat heart-shaped

subdenta'tus: nearly toothless

subdivarica'tus: slightly spreading at a wide angle

subercula'tus: of cork; corky

suberec'tus: somewhat erect

subero'sus: cork-barked →

subfalca'tus: somewhat curved
   or sickle-shaped

subglau'cus: somewhat covered
   with a fine white powder

subhirtel'lus: somewhat hairy

subluna'tus: somewhat crescent-shaped

submer'sus: submerged

subperen'nis: nearly perennial

subpetiola'tus: partially petioled or leaf-stalked

subscan'dens: partially climbing

subses'silis: almost stalkless

subsinua'tus: somewhat sinuate; with indented margins
   between lobes or divisions

subterra'neus: underground

subula'tus: awl-shaped

subumbella'tus: somewhat umbellate

subvillo'sus: with rather soft hairs

subvolu'bilis: somewhat twining

succotri'nus: from the island of Socotra

succulen'tus: succulent; fleshy; juicy

BLOSSOM TEA

Subhirtellus. In *Gardens in Winter,* Elizabeth Lawrence recounts this legend about the discovery of the autumn-flowering cherry, *Prunus subhirtella* var. *autumnalis:* "Once on a warm November afternoon in the fifth century, as the Emperor of Japan was taking tea in his garden, a cherry petal drifted into his cup. He immediately sent his servants in search of the tree it had fallen from and they were not to come back until they found it."

LILIES OF THE FIELD
Superbiens, superbus. The turk's-cap lily, *Lilium superbum,* is indeed a superb example of a wildflower in all its glory. Orange or scarlet, with purple spots within, they can be found in clusters of up to forty radiant blossoms and may grow to a height of eight feet. Lilies have been admired and praised down through the ages. "Consider the lilies of the field, how they grow; ∾ They toil not, neither do they spin; ∾ And yet I say unto you, that even Solomon in all his glory ∾ Was not arrayed like one of these."
Matthew 6:28–29

suec'icus: Swedish

suffrutes'cens, suffrutico'sus: somewhat shrubby

sufful'tus: supported

sulca'tus: sulcate; grooved; furrowed lengthwise

sulphu'reus: sulfur-colored

sumatra'nus: of Sumatra, an Indonesian island

super'biens, super'bus: superb; proud

supercilia'ris: eyebrow-like

super'fluus: superfluous; redundant

supi'nus: prostrate; flat on the ground

supraaxilla'ris: above the axils

supraca'nus: with grayish hairs above

surculo'sus: producing suckers

susia'nus: of Susa, an ancient city of Persia →

suspen'sus: suspended; hanging

sylvat'icus: belonging to woods or forests

sylves'ter, sylves'tris: of woods or forests

sylvic'olus: growing in woods

syphilit'icus: syphilitic; of, caused by, or having syphilis; formerly used in the treatment of syphilis

syri'acus: Syrian

syringan'thus: with flowers like those of the lilac, *Syringa*

syringifo'lius: with leaves like those of the lilac, *Syringa abulaefor'mis, tabulaifor'mis:* flat; resembling a table or board

tabula'ris: table-like; flattened horizontally

taedig'era: cone-bearing; torch-bearing

tanacetifo'lius: tansy-leaved; with leaves like *Tanacetum*

taraxicifo'lius: dandelion-leaved; with leaves like *Taraxicum*

tardiflo'rus: late-flowering

tardi'vus: tardy; late

tarta'reus: with a loose or rough, crumbling surface

tatar'icus: from Tartary (the portion of Siberia and Russia formerly inhabited by Tatars)

tau'reus: of oxen

tau'ricus: Taurian; Crimean

tauri'nus: bull-like; pertaining to cattle

taxifo'lius: yew-leaved; with leaves like those of the yew tree, *Taxus* →

tech'nicus: technical; special

tecto'rum: of roofs or houses

tec'tus: concealed; covered

tellimoi'des: resembling *Tellima*

temulen'tus: drunken

tenacis'simus: most tenacious

te'nax: tenacious; strong

SPRING GREENS
Taraxicifolius. Dandelion salad for four. Trim and wash 1 lb. dandelion greens. Slice 2 ripe tomatoes. Thinly slice two scallions. Mix 4 Tbs. vinegar, 1 ½ tsp. sugar, 1 clove garlic (minced), ¼ tsp. freshly ground pepper. Fry and crumble 6 slices of bacon. Away from the fire add greens to hot bacon fat; stir until they wilt a little. Toss in the vinegar mixture. Arrange tomato slices on a plate, mound dandelion greens on each, top with bacon and scallions.

❦

BUTTERFLY FOOD
Tataricus. Aster tataricus hails from Siberia. Its pale violet flowers provide a nectar that fuels millions of monarch butterflies on their incredible annual journey to El Rosario, Mexico, to nest.

A TENDER BEAUTY
Tenera. Despite its tropical origins and aptness of its species name, *Verbena tenera* var. *maonettii* deserves a place in every sunny garden. From May to October its low-growing mounds of foliage are covered with clusters of flowers edged in white. While it is only reliably hardy, it is easy to root from cuttings taken in early fall.

tenebro'sus: of dark or shaded places

tenel'lus: slender; tender; soft

te'ner, ten'era: slender; tender; soft

tentacula'tus: with tentacles

tenuicau'lis: slender-stemmed

tenuiflo'rus: slender-flowered

tenuifo'lius: slender-leaved ←

tenuil'obus: slender-lobed

tenu'ior: more slender

tenuipet'alus: slender-petaled

ten'uis, tenuis'simus: slender; thin

tenuisty'lus: slender-styled

terebintha'ceus: of turpentine

terebinthifo'lius: Terebinthus-leaved; with leaves smelling like turpentine

te'res: terete; circular in cross-section; cylindrical

teretifo'lius: terete-leaved

tereticor'nis: with terete or cylindrical horns

termina'lis: terminal

ternate'a: of Ternate Island (in the Moluccas, Indonesia)

terna'tus: in threes

ternifo'lius: with leaves in threes

terres'tris: of the earth

tessella'tus: tessellate; checkered

testa'ceus, testa'ceous: light brown; brick-colored

testicula'tus: testiculated; resembling testicles

testudina'rius: like a tortoiseshell

tetracan'thus: four-spined

tetragono'lobus: with a four-angled pod

tetram'erus: of four members

tetran'drus: four-anthered

tetran'thus: four-flowered

tetraphyl'lus: four-leaved

tetrap'terus: four-winged

tetraque'trus: four-cornered

teucrioi'des: resembling the herb, *Teucrium*

tex'anus: of Texas; Texan

tex'tilis: textile; woven

thalictroi'des: resembling the meadow rue, *Thalictrum*

thapsoi'des: resembling the mullein, *Verbascum thapsus*

theba'icus: of Thebes (now in Greece)

theif'era: tea-bearing

therma'lis: warm; of warm springs

thibet'icus: of Tibet; Tibetan

thurif'era: incense-bearing

thuyoi'des, thyoi'des: resembling *Thuja*

thymifo'lius: thyme-leaved

thymoi'des: thyme-like

thyrsiflo'rus: having flowers in a thyrse (a dense cluster around the flower's axis)

FOR STRESS AND STIES ↑
Thapsoides. Verbascum thapsus—or mullein— has many medicinal uses. To still a painful cough, the flowers of the plant may be used to make a soothing tea. Topically, the same solution is an effective remedy for eye infections and sties.

thyrsoi'des: thyrse-like

tibet'icus: of Tibet; Tibetan

tibic'inis: flute-like

tigri'nus: striped like a tiger

tilia'ceus: Tilia-like; linden-like; resembling the linden, *Tilia* ←

tiliaefo'ius: with foliage like that of *Tilia*

tincto'rius: used for dyes; pertaining to dyes or dyers

tinc'tus: dyed

tingita'nus: of Tangier, Morocco

tipulifor'mis: resembling a daddy longlegs in shape

tita'nus: very large

tomento'sus: tomentose; densely wooly

ton'sus: clipped; sheared

tormina'lis: useful against colic

toro'sus: cylindrical with contractions at intervals

tortifo'lius: with twisted leaves

tor'tilis: twisted

tortuo'sus: very twisted

tor'tus: twisted

torulo'sus: somewhat torose; see *torosus*

toxica'rius, tox'icus: poisonous

toxif'era: poison-producing

trachypleu'ra: rough-ribbed or -nerved

trachysper'mus: rough-seeded

tragophyl'lus: having leaves with a goat-like odor

TINCTURES
Tinctorius. Many plants are relied upon by weavers and dye specialists. Blue coloring can be extracted from the leaves of the *Isatis tinctoria* and the brightly colored roots of *Rubia tinctorum* are used to create a vivid red dye.

translu'cens: translucent

transpa'rens: transparent

transylvan'icus: of Transylvania

trapezifor'mis: with four unequal sides

trapezioi'des: trapezium-like; with four sides, none parallel

tremuloi'des: trembling; quaking; shaking ↓

trem'ulus: quivering; trembling

triacanthoph'orus: bearing three spines

triacan'thus: three-spined

trian'drus: with three anthers or stamens

triangula'ris, trianagula'tus, trian'gulus: three-angled

tricauda'tus: three-tailed

triceph'alus: three-headed

tricho'calyx: hairy-calyxed

trichocar'pus: hairy-fruited

trichomanefo'lius: having leaves like those of *Trichomanes*

trichomanoi'des: resembling *Trichomanes*

trichophyl'lus: hairy-leaved

trichosan'thus: hairy-flowered

trichosper'mus: hairy-seed·ied

trichot'omus: three-branched or three-forked

tricoc'cus: three-seeded; three-berried

tri'color: three-colored

tricor'nis: three-horned

TRANSFERENCE
Tremuloides. The leaves of the quaking aspen, *Populus tremuloides,* tremble and shiver in even the slightest breeze. Country people believed the shaking could cure their fever and chills. Those who were ailing would cut a lock of their hair, tie it to the tree, and say: "Aspen tree, aspen tree, ∾ Shake and shiver instead of me."

❧

MANY-SPLENDORED
Trichotomus. In July the Japanese glory-bower, *Clerodendrum trichotomum,* produces clusters of sweetly fragrant white flowers surrounded by rosy calyxes. Its three-branched corolla and petals drop in late summer, and turquoise berries form within the calyxes, ripening to a deep blue. Usually grown as a shrub, it is fairly hardy.

GREEN INSULATION
Tricuspidatus. Parthenocissus tricuspidata, a vine with three-pointed leaves, is commonly known as Boston ivy. Trained to grow on one side of a house, the large leaves and dense foliage of Boston ivy provide a cooler room inside. When the leaves fall in winter, the sunlight filters through the vine and contributes to the warmth inside.

❦

A HELLISH HEDGE
Trifoliatus. Three-leaved, thorny-branched, the trifoliate orange, *Poncirus trifoliata,* is often used as a loose hedge. With trimming, it shapes out neatly, bearing little white flowers in spring, and golden fruits in the fall.

❦

tricuspida'tus: having three points

tridac'tylus: three-fingered

tri'dens, tridenta'tus: three-toothed

trifascia'tus: three-banded

trif'idus: three-parted

triflo'rus: three-flowered

trifolia'tus: three-leaved →

trifoliola'tus: with three leaflets

trifo'lius: three-leaved

trifurca'tus, trifur'cus: three-forked

triglochidia'tus: with three barbed bristles

trigonophyl'lus: triangular-leaved

trilinea'tus: three-lined

triloba'tus, tril'obus: three-lobed

trimes'tris: of three months

triner'vis: three-nerved

trinota'tus: three-marked or three-spotted

triornithoph'orus: bearing three birds

triparti'tus: three-parted

tripet'alus: three-petaled

triphyl'lus: three-leaved

trip'terus: three-winged

tripuncta'tus: three-spotted

trique'tris: three-cornered

trisper'mus: three-seeded

trista'chyus: three-spiked, refers to the flower spike

tris'tis: sad; bitter; dull

triterna'tus: thrice; in threes

trium'phans: triumphant

trivia'lis: common; ordinary

trolliifo'lius: resembling the leaves of *Trollius*, globe flower

trop'icus: of the tropics

truncat'ulus: somewhat truncated

trunca'tus: truncated; squared-off

tubaefor'mis: trumpet-shaped

tuba'tus: trumpet-shaped

tubercula'tus, tuberculo'sus: having tubercles (nodules)

tubero'sus: tuberous →

tubif'era: tube-bearing

tubiflo'rus: trumpet-flowered

tub'ispathus: tube-spathed

tubulo'sus: with tubes

tulipif'era: tulip-bearing

tu'midus: swollen

turbina'tus: top-shaped

turbinel'lus: small and top-shaped

tur'gidus: turgid; inflated; full

typhi'nus: pertaining to fever

typ'icus: typical

SANS SERMON
Triphyllus. Arisaema triphyllum has three leaves and a tall stalk with a flower that tilts, forming a cover for the green and brown "pulpit" underneath. A small protuberance inside the pulpit is "Jack," thus the name jack-in-the-pulpit.

UPON INTRODUCTION TO SOCIETY
Tuberosus. Helianthus tuberosus, the Jerusalem artichoke or Native American sunflower, grows from knobby, white-fleshed tubers. Growing to a height of twelve feet, *H. tuberosus* produces butter yellow flowers in late summer and fall. After frost cuts the foliage to the ground, the tubers can be dug and eaten. Frost brings out the sweet, nut-like flavor of the tubers. They can be eaten raw or cooked.

INFLATE & GIBBOUS

Urceolatus. "The Corolla is *Urceolate,* Pitcher-shaped, when it is inflate and gibbous on all Sides, after the Manner of that Vessel."
James Lee, *An Introduction to Botany,* 1765

❧

BUT INEDIBLE

Usneoides. Tillandsia usneoides, Spanish moss, gets its specific name from its resemblance to the lichen *Usnea,* which also grows in trees. Not parasitic to the trees it drapes, Spanish moss can photosynthesize and obtain water through tiny scales on its stems. Despite its common name it is not a moss but a member of the *Bromeliad* family, which also includes pineapples.

❧

lic'inus: resembling *Ulex,* gorse

uligino'sus: of wet or marshy places

ulmifo'lius: with leaves like the elm, *Ulmus*

ulmoi'des: resembling the elm, *Ulmus*

umbella'tus: with umbels or umbrella-like flower clusters ↓

umbellula'tus: with umbellets or clusters

umbona'tus: bearing at center an umbo (a knob or stout projection, as on a shield)

umbraculif'era: umbrella-bearing

umbro'sus: shaded; shade-loving

uncina'tus: hooked at the point

unda'tus: wave-like; wavy

undulatifo'lius: wavy-leaved

undula'tus: undulated; wavy

undulifo'lius: wavy-leaved; with undulate foliage

unguicula'ris, unguicula'tus: narrow-clawed (refers to the base of the petals)

unguipet'alus: with claw-shaped petals

unguispi'nus: with hooked spines

uni'color: one-colored

unicor'nis: one-horned

unidenta'tus: one-toothed

uniflo'rus: one-flowered

unifo'lius: one-leaved

unilatera'lis: one-sided

unioloi'des: resembling *Uniola*, sea oats

univitta'tus: one-striped

urba'nus: city-loving; urban

urceola'tus: urn-shaped

u'rens: burning; stinging

urentis'simus: causing extreme burning or stinging

urnig'era: pitcher-bearing

urophyl'lus: with leaves resembling a tail

urosta'chyus: with spikes resembling tails

ursi'nus: pertaining to bears; northern (under the
constellation of the Great Bear)

urticaefo'lius, urticifo'lius: nettle-leaved →

urticoi'des: nettle-like

usitatis'simus: most useful

usneoi'des: resembling *Usnea*, a lichen that looks like Spanish
moss, *Tillandsia usneoides*

ustula'tus: burnt; sere; dried-up; withered

u'tilis: useful

utilis'simus: most useful

utricula'tus, utriculo'sus: having a small bladder or one-
seeded fruit (utricle)

uvif'era: bearing grapes

BLADDERS ARE
UTRICLES
Utriculatus. Leaves with
floating bladders give
the *Utricularia*, or blad-
derworts, their name.
"The slender *Utricularia*,
a dainty maiden whose
light feet scarce touch
the water."
T. W. Higginson,
Out-Door Papers, 1863

FROM ONE ACORN
Velutinus. The black oak, *Quercus velutina*, is so named because its buds are covered with a velvety pale yellowish gray pubescence. Its inner bark is yellow or bright orange and was once a source of tannin and yellow dye. The oak apple and leaves were once worn as a symbol of loyalty to royalty. Now the U.S. military awards the bronze or silver oak leaf cluster for heroism.

FOR SNAKEBITE?
Venosus. Hieracium venosum, rattlesnake-weed. Leaves with distinctive purple veins give this wildflower its specific name. Its common name arises from the belief that it is a useful treatment for snakebite.

vaccinifo'lius: with leaves like those of the blueberry, *Vaccinium*

vaccinoi'des: resembling blueberry, *Vaccinium* ↓

vacil'lans: swaying; inconstant

va'gans: wandering

vagina'lis, vagina'tus: sheathed

valdivia'nus: of Valdivia, Chile

valenti'nus: of Valencia, Spain

val'idus: strong

vanda'rum: of *Vanda* (an orchid)

varia'bilis, va'rians, varia'tus: variable

varico'sus: varicose; abnormally swollen or dilated

variega'tus: variegated

variifo'lius: variable-leaved

variifor'mis: of variable forms

va'rius: various; diverse

vegeta'tus, veg'etus: vigorous

vela'ris: pertaining to curtains or veils

ve'lox: rapidly growing; swift

velu'tinus: velvety

venena'tus: poisonous

veno'sus: veiny

ventrico'sus: ventricose

venus'tus: handsome; charming

verbascifo'lius: with leaves that look like mullein, *Verbascum* →

verecun'dus: modest; blushing

vermicula'tus: worm-like

verna'lis: of spring

vernicif'era, vernicif'lua: varnish-bearing

vernico'sus: varnished

ver'nus: of spring

verruco'sus: verrucose; covered with warts

verruculo'sus: very warty

versi'color: variously colored

verticilla'ris, verticilla'tus: verticillate; whorled

ve'rus: true; genuine; standard

ves'cus: weak; thin; feeble

vesiculo'sus: having little bladder-like structures

vesperti'nus: of the evening; western

vesti'tus: covered; clothed

vex'ans: puzzling; vexatious; irksome

vexilla'rius: with a standard upright petal

viburnifo'lius: with leaves like those of *Viburnum*

viciaefo'lius, vicifo'lius: with leaves like those of vetch, *Vicia*

victoria'lis: victorious

villo'sus: villous; soft-haired

vimina'lis, vimin'eus: with long slender shoots

A PRINCESS OF A PEA
Villosus. The Carolina bush pea, *Thermopsis villosa*, grows from four to five feet tall and flowers in May. Its yellow flowers are followed by fuzzy gray seed pods often used in dried arrangements.

❧

FOR EGGS AND PORCHES
Viminalis. The long, slender twigs of the basket willow (*Salix viminalis*) were used by the pioneers for basketry and wickerwork.

❧

VISCOUSNESS ↑
Viscosus. The young twigs, leaf stalks, flower stalks, and pods of the clammy locust (*Robinia viscosa*) are covered with gland hairs that secrete a sticky—or clammy—substance.

❦

SPRING CLEANING
Vomitorius. The evergreen leaves and red berries of the yaupon (*Ilex vomitoria*) make its twigs prized Christmas decorations. Native Americans used the leaves to prepare a vomit-inducing tea. Tribes from the interior traveled to the coast each spring to partake of this tonic, a sort of spring-cleaning ritual.

vinif'era: wine-bearing; grape-bearing

vino'sus: wine red in color

viola'ceus: violet-colored

violes'cens: becoming violet; light violet

vi'rens: green

vires'cens: almost green; light green

virga'tus: twiggy

virgina'lis, virgin'eus: virginal; white

virginia'nus, virgin'icus, virginien'sis: from Virginia, United States

virides'cens: almost green; pale green

viridicarina'tus: green-keeled

viridiflo'rus: green-flowered

viridifo'lius: green-leaved

viridifus'cus: green-brown

vir'idis: green

viridis'simus: very green

virid'ulus: greenish

viscid'ulus: somewhat sticky

vis'cidus: viscid; sticky; syrupy; thick

viscosis'simus: very sticky

visco'sus: sticky

vita'ceus: grape-like; resembling *Vitis*, grapes ↗

vitelli'nus: having the color of an egg yolk

viticulo'sus: sarmentose; producing long, winding runners or stolons

vitifo'lius: grape-leaved

vitta'tus: striped lengthwise

vittig'era: bearing stripes

vivip'arus: bearing live young; self-propagating by production of plantlets

volgar'icus: of the Volga River, eastern Russia

volu'bilis: twining

volu'tus: with rolled leaves

vomito'rius: causing vomiting; emetic

vulcan'icus: of or pertaining to Vulcan, the Roman god of fire and metalworking; growing on a volcano

vulga'ris, vulga'tus: vulgar; common

vulpi'nus: relating to foxes; used for species with inferior fruit

Aloe vera vulgaris.

ANCIENT LAXATIVE ↑
Vulgaris. Tanacetum vulgare, common tansy, was brought to America by English settlers. Symbolic of the bitter herbs of the Paschal Feast, tansy was used to flavor a traditional cake or pudding served on Easter. Tansy was thought to have a cleansing effect on the system after Lenten fasts, but beware, an overdose of tansy can cause death.

✿

A PLANT FOR MOST
SEASONS
Xanthorhizus. Xanthorhiza simplicissima is a deciduous, vertically growing shrub. It is grown not only for its small star-shaped flowers, which blossom in the spring, but for its striking foliage—bright green ovals that turn bronze or purple in autumn. It spreads vegetatively by yellow stems that grow underground, whereby it acquired its name, meaning yellow-rooted.

xanthacan'thus: yellow-spined
xanth'inus: yellow
xanthocar'pus: yellow-fruited
xantholeu'cus: yellowish white
xanthoneu'rus: yellow-nerved
xanthophyl'lus: with yellow leaves
xanthorhi'zus: yellow-rooted
xantho'xylon: with yellow heartwood ↓
xylocan'thus: woody-spined

Zebri'nus: zebra-striped

zeylan'icus: of Ceylon, an island off the southern tip of India, now Sri Lanka

zibethi'nus: smelling like a civet cat; foul-smelling

zizanioi'des: resembling Zizania, a wild rice

zona'lis, zona'tus: zoned or banded with a distinct color

FOR MALINGERERS
Zonalis, zonatus.
The *Crocus zonatus* is a golden-throated, lilac-colored flower that blooms in early autumn. It was once thought that autumn-blooming crocuses were poisonous while their spring-blooming counterparts were benign. *Crocus colchicum,* a toxic flower, was popular in ancient Greece among malingering slaves. They would consume just enough of the bulb to make themselves too ill to work the next day.

Bailey, L. H. *Hortus Third.*
New York: Macmillan Publishing Company, Inc., 1976.

Bailey, L. H. *How Plants Get Their Names.*
New York: Dover Publications, 1963.

Borror, Donald J. *Dictionary of Word Roots and Combining Form.*
Palo Alto: National Press Book, 1971.

Coombes, Allen J. *Dictionary of Plant Names.*
Portland: Timber Press, 1985.

Gledhill, D. *The Names of Plants.*
New York: Cambridge University Press, 1989.

Healy, B. J. *A Gardener's Guide to Plant Names.*
New York: Charles Scribner's Sons, 1972.

Johnson, A. J., and H. A. Smith. *Plant Names Simplified.*
London: Landsman Bookshop, 1979.

Oxford English Dictionary. Oxford: Clarendon Press, 1990.

Oxford Latin Dictionary. Oxford: Clarendon Press, 1982.

Smith A. W. *A Gardener's Book of Plant Names.*
New York: Harper and Row Publishers, 1963.

Stearn, William T. *Botanical Latin.*
London: David and Charles, 1973.